Lighting Design in Shared Public Spaces

This book advocates an approach to lighting design that focuses on how people *experience* illumination. *Lighting Design in Shared Public Spaces* contextualises light, dark and lighting design within the settings, sensations, ideas and imaginaries that form our understandings of ourselves and the world around us.

The chapters in this collection bring a new perspective to lighting design, arguing for an approach that addresses how lighting is experienced, understood and valued by people. Across a range of new case studies from Australia, Germany, Denmark, and the United Kingdom, the authors account for lighting design's crucial role in shaping our dynamic and messy experiential worlds. With many turning to innovative ethnographic methodologies, they powerfully demonstrate how feelings of comfort, safety, security, vulnerability, care and well-being can configure in and through how people experience and manipulate light and dark. By focusing on how lighting is improvised, arranged, avoided and composed in relation to the people and things it acts upon, the book advances understandings of lighting design by showing how improved experiences of the built environment can result from more sensitive and context-specific illumination.

The book is intended for social scientists who are interested in the lit or sensory world, as well as designers, architects, urban planners and others concerned with how the experience of light, dark and lighting might be both better understood and implemented in our shared public spaces.

Shanti Sumartojo is Associate Professor of Design Research in the Faculty of Art, Design and Architecture, and a member of the Emerging Technologies Research Lab at Monash University. Grounded in human geography, and with a strong commitment to interdisciplinary and collaborative scholarship, her research includes theoretically informed inquiry into how people experience design and technology in their surroundings, particularly in shared, public spaces. Her recent books include *Atmospheres and the Experiential World* (2018) and *Geographies of Commemoration in a Digital World* (2021).

Lighting Design in Shared Public Spaces

Edited by Shanti Sumartojo

Routledge
Taylor & Francis Group

NEW YORK AND LONDON

First published 2022
by Routledge
605 Third Avenue, New York, NY 10158

and by Routledge
2 Park Square, Milton Park, Abingdon, Oxon, OX14 4RN

Routledge is an imprint of the Taylor & Francis Group, an informa business

Library of Congress Cataloging-in-Publication Data
Names: Sumartojo, Shanti, editor.
Title: Lighting design in shared public spaces / edited by Shanti Sumartojo.
Description: New York, NY : Routledge, 2022. | Includes bibliographical references and index.
Identifiers: LCCN 2021046842 (print) | LCCN 2021046843 (ebook) | ISBN 9781032022642 (hardback) | ISBN 9781032022635 (paperback) | ISBN 9781003182610 (ebook)
Subjects: LCSH: Municipal lighting. | Lighting, Architectural and decorative. | Aesthetics.
Classification: LCC TK4188 .L538 2022 (print) | LCC TK4188 (ebook) | DDC 628.9/5—dc23
LC record available at https://lccn.loc.gov/2021046842
LC ebook record available at https://lccn.loc.gov/2021046843

ISBN: 978-1-032-02264-2 (hbk)
ISBN: 978-1-032-02263-5 (pbk)
ISBN: 978-1-003-18261-0 (ebk)

DOI: 10.4324/9781003182610

Typeset in Univers
by Apex CoVantage, LLC

Contents

Contents

Figures

Figures

Contributors

Jess Berry is Senior Lecturer in the Department of Design and Researcher of the Monash University XYX Gender + Place research lab. Her research focuses on how gender identities are articulated and mediated through and by space. She is a contributing co-editor of *Contentious Cities: Design and the Gendered Production of Space* (Routledge 2020) and the author of *House of Fashion: Haute Couture and the Modern Interior* (Bloomsbury 2018) as well as the forthcoming monograph *Cinematic Style: Fashion, Architecture and Interior Design on Screen* (Bloomsbury 2022).

Mikkel Bille holds a PhD in Social Anthropology from University College London and is Associate Professor at Roskilde University. His research focuses broadly on the role of material culture in our everyday lives, previously focused on the Middle East and now predominantly in Scandinavia. He has a long interest in exploring how we use elements as intangible as light and luminosity to make sense of our surroundings in diverse ways. He is currently leading a project on Nordic lighting, which investigates how new technologies have been implemented in Scandinavia to stage urban spaces and shape the feelings of cities.

Melisa Duque is Research Fellow in the Emerging Technologies Research Lab and Full-Time Member of the Department of Design at Monash University. As a design researcher, her work sits at the intersection of design anthropology, participatory design and everyday design.

Casper Laing Ebbensgaard is Lecturer in Cultural Geography at University of East Anglia. His research explores the affective and aesthetic politics of architectural design and urban change with a particular interest in the geographies of verticality, home and belonging, and the urban night. He is Co-founder of the research collaborative *The Urban Night Project* (urbannightproject.com) and Curator of the exhibition *Midnight Sun* (2021) at Black Tower Projects, which explored how the vertical city is experienced, imagined and sustained at night. He recently completed a Leverhulme Early Career Fellowship.

Tim Edensor is Professor of Human Geography at Manchester Metropolitan University and Principal Research Fellow at Melbourne University. He is the author of *Tourists at the Taj* (1998); *National Identity, Popular Culture and Everyday Life* (2002); *Industrial Ruins: Space, Aesthetics and Materiality* (2005); *From Light to Dark: Daylight, Illumination and Gloom* (2017) and *Stone: Stories of Urban Materiality* (2020). He also edited *Geographies of Rhythm* (2010). Tim has written extensively on national identity, tourism, ruins, mobilities and landscapes of illumination and darkness.

Ellen Kathrine Hansen, PhD, is Associate Professor, and Co-founder and Head of the MSc program and Lighting Design Research Group at Aalborg University, Copenhagen. Hansen holds a Master in Architecture from the Royal Academy of Fine Arts in Copenhagen and a PhD in transdisciplinary Lighting Design processes from Aalborg University. She has more than 25 years of experience driving industrial and academic projects developing new sustainable and architectural potentials through an integration of daylight, lighting and responsive technology.

Mette Hvass holds a Master in Architecture from the Royal Academy of Fine Arts in Copenhagen. She has 15 years practice experience as an architect and lighting designer and is currently PhD Student in Lighting Design at the Department of Architecture, Design and Media Technology at Aalborg University in Copenhagen. Her research focuses on the sensory experience of lighting in the urban context and how lighting levels influence connectedness to space, surroundings and people.

Olivia Norma Jørgensen is Master Student at the Department of Anthropology, University of Copenhagen. She is Student Helper at the *Living with Nordic Lighting* project at Roskilde University, where she has done fieldwork on squares around Copenhagen. She has a special interest in applied anthropology and in how global companies use social science to create growth in their business.

Nicole Kalms is Associate Professor in the Faculty of Art Design and Architecture and Founding Director of the Monash University XYX Lab, which leads national and international research in Gender and Place. Kalms' research examines digital, experiential, political and material interventions to articulate both the shared and conflicted struggles of women and girls internationally. Her praxis repositions design as a strategic tool for challenging gender inequity. Her recent research has focused on public transport spaces for women and girls, gender-sensitive CPTED and the use of participatory co-design to challenge gender-neutral urban policy.

Stine Louring Nielsen is Design Anthropologist and PhD in Lighting Design. For over a decade, she has been working together with architects, designers and users of healthcare environments on projects linking aesthetics, health and well-being. Her research focuses on the potential of art, colour, lighting, healthcare designs and healing architecture to attune bodily sensations, experiences and practices. Stine currently applies her insights into the intersections of the build environment and the human body at the Bevica Foundation, which conveys and supports interdisciplinary research projects linking universal design and the Leave No One Behind agenda of the UN's Sustainable Development Goals.

Sarah Pink is Professor, Design Anthropologist and Director of the Emerging Technologies Research Lab at Monash University. Sarah is known globally for her leadership in innovative digital, visual and sensory research and dissemination methodologies. She engages these methodologies – via interdisciplinary projects in design, engineering and creative practice – to engage with contemporary issues and challenges. Her current projects investigate future mobilities and digital energy futures. Sarah's recent books include *Imagining Personal Data* (2019), *Atmospheres and the Experiential World* (2018), *Uncertainty and Possibility* (2018), *Making Homes* (2017) and *Anthropologies and Futures* (2017).

Nona Schulte-Römer is currently Guest Professor and Researcher at the Institute for European Ethnology at Humboldt University Berlin. She has studied city lights since 2008, when she started her sociological inquiry into the LED 'revolution' in urban public lighting

(*Innovating in Public*, TU Berlin, 2015). In her current light-related research, she explores the public understanding of light pollution and sustainable LED lighting. She is co-author of the free e-publication *Light Pollution – A Global Discussion* (www.ufz.de/light-pollution), and part of a citizen science network that documents light sources emissions in public spaces (nachtlicht-buehne.de/nachtlichter).

Laurene Vaughan is Professor and Dean of the School of Design at RMIT University. She is a practicing artist, designer and educator, who, through her research, explores and comments on the interactive and situated nature of human experience.

Karen Waltorp, PhD, is Associate Professor at Department of Anthropology, University of Copenhagen, and Head of Ethnographic Exploratory, a physical and virtual space for multimodal experimentation across faculty, students and guests. Waltorp is interested in people and their (digital) environments, and is Author of *Why Muslim Women and Smartphones: Mirror Images* (Routledge, 2020), Coordinator of Researcher Group Technē: and Co-coordinator of the Future Anthropologies Network under the European Association of Social Anthropologists.

Hoa Yang is PhD Candidate at the Monash University Emerging Technologies Lab and practising Senior Lighting Designer at Arup. Hoa specialises in daylight and experiential lighting design in her role at Arup and has extensive experience in local and international arts and culture, public infrastructure, and urban realm projects. Hoa's practice brings together 24-hour lighting design with human-centric strategies to transform experiences for minority groups within the built environment. Her current research focuses on the role of technical lighting parameters for positive night-time experiences for intersectional people in public spaces.

Chapter 1

LIGHT, DARK AND LIGHTING DESIGN FOR SHARED PUBLIC SPACES

New perspectives on experiences of the lit world

Shanti Sumartojo

DOI: 10.4324/9781003182610-1

Introduction

Light is central to how we experience our worlds. It shapes how we feel about and are affected by our surroundings, what we are able to do in them, and how we relate to the people, things, materials and spaces around us. It is central to how we imagine, remember and anticipate our lives and, as such, is far more than a means for visual apprehension; indeed, 'the perception of luminous and gloomy space is a key existential dimension of living in the world, of the experience of space and time' (Edensor 2017: vii; see also Pink and Sumartojo 2018).

If light profoundly affects how we feel, then this means that lighting design is also bound up in and responsible for the distinctive ways that these feelings emerge as we inhabit or move through the places subject to the practice of lighting design. Accordingly, this book advocates for an approach to lighting design that focuses on how people *experience* illumination: how it feels, what it makes possible or imaginable and what it shuts down or limits. Across a range of new work from Australia, Germany, Denmark, and the UK, the authors seek to account for complex, dynamic and messy experiential worlds, and to directly address lighting design's crucial role in shaping these. The book deliberately draws together authors from a range of fields and, as such, is intended for social scientists who are interested in the lit or sensory world, but also designers, architects and others interested in how the experience of light and lighting might be both better understood and implemented in professional practice.

Taken together, the chapters in the book powerfully demonstrate how feelings of comfort, safety, security, vulnerability, care and well-being can configure in and through how people experience and manipulate light, and that improved experiences of the built environment can result from more sensitive and context-specific lighting design. As such, it builds on existing research in lighting design that seeks to understand the feelings, sensations, impressions and thoughts of the users of technology as they interact with it (Simonsen 2018; Jensen et al. 2018). However, while this large and varied body of work has developed robust methodologies, much of it occurs in lab-based settings where variables can be controlled and aspects of technologies rigorously tested. Some lighting design research has addressed responses to or interactions with technology 'in the wild', but even here, in the main, research focuses on lighting technologies themselves, implementing well-established research processes focused on specific improvements.

In contrast, we advance a research approach that contextualises this technology in the dynamic and emergent everyday worlds of people, and that considers how we understand and make use of technology as part of everything else we do. Although lighting designers require specific insights into technologies as part of design processes, we argue that the spatial, social and experiential contexts of light are vital in terms of how people address, make use of and modify or improvise with technologies to get the best from them. As such, the book treats lighting design as agential in helping to configure an experiential lit world in which we all dwell and move, and many of the chapters grapple with this by adopting an ethnographic approach to investigating light and lighting.

This means that we tend to move away from defining people as 'users' of lighting, arguing that such a perspective casts people in a particular type of relationship with technology. To speak of 'users' defines design not in the terms that people themselves may use to understand it, but in terms that speak to the designers' intentions and goals. While this is valuable in terms of design development and testing processes, it can risk overlooking important aspects that are not directly related to people but instead are part of their environments, which often include other technologies (including, as discussed in this book, light sources).

This book, therefore, adopts an expansive stance in terms of light and lighting technology by approaching it as it exists in relation to other things, feelings, ideas and processes already in place and that form part of how people understand themselves and the world around them. Across its ten chapters, this book focuses on many different aspects of light and lighting design as it is part of our everyday lives. This is because light – which by its nature can seep, infuse, dissipate or mingle with other aspects of our surroundings – is part of the messy and ongoing nature of experience, its moods and atmospheres, and its material, sensory and affective qualities (Jensen et al. 2018). The chapters bring social scientific concepts, research methods and critical analysis together with design methods that focus on lighting use and manipulation, thus opening the way to valuable new insights. The book focuses on how lighting is experienced and understood but also improvised, arranged, avoided and composed in relation to the people and things proximate to it. While we hope that the

Figure 1.1
The complexity of the lit world: a small boat moves across the surface of the Yarra River in Melbourne, an environment enlivened by the sparkle of light reflected in the water, the illumination of surrounding buildings and the pilot lights of the boat itself.
Photo: Shanti Sumartojo.

book can help improve understanding of lighting design and its use in built environments, it is also intended to help frame the possibilities and limits of lighting by showing how it relates to people's lives, activities and feelings.

In the rest of this introduction, I draw together recent research on experience in lighting design with scholarship on light in anthropology and human geography, fields with a shared interest in the experiential world. I develop this by way of an important concept that informs this book and that underpins the methodologies that appear in it: the 'lit world' (Pink and Sumartojo 2018). Following this, I explain the structure and various contributions of the book's chapters.

Accounting for the experiential world in lighting design research

Research directed at understanding the perceptions and responses of people to designed objects or processes often aims to understand the effects of these design artefacts and, ultimately, improve their performance for an intended function. However, such a focus on discrete aspects of interaction between people and technologies, objects, or processes often requires the partitioning of human experience into measurable or describable units of analysis, such as aesthetics, emotion or ergonomics; indeed, researchers themselves sometimes question the emphasis on measurement of ambiguous or ineffable qualities of feeling (Simonsen 2018: 199) or raise ethical concerns about the uses of such approaches to support the potential manipulation of people (Gray et al. 2018). As a result, some studies of how people use or experience design and technology have turned to interdisciplinary approaches that use qualitative, ethnographic techniques beyond controlled or laboratory-based conditions (e.g., Jensen et al. 2018).

In studies of lighting technologies, however, many well-established methods evaluate those technologies in lab-based settings. The majority of such studies place those technologies at the centre of inquiry, focusing on their affordances and how they interact with their surroundings, and assessing aspects of how research participants perceive these. This includes a small but growing focus on how spatialised 'atmospheres' relate to lighting design (Kuijsters et al. 2015; Stokkermans et al. 2017, 2018; Albertazzi et al. 2018; Vogels 2008). For example, Stokkermans et al. (2017, 2018) address the atmosphere or the 'impression' of a lit space with experiments that varied the lighting conditions in a controlled environment and assessed people's perceptions of those spaces. Setting out to understand how 'light designs translate into the ultimate experience of a space', they identified the importance of the 'intensity, colour, beam shape and position of the light sources' (Stokkermans 2017: 1164) and the importance or the way people perceived these as well as the location they were placed. This built, in part, on earlier work (Vogels 2008) that sought to 'quantify the atmosphere of an environment'. Defining atmosphere as an 'affective evaluation of an environment', Vogels (2008) identified the key measures of cosiness, liveliness, tenseness and detachment – atmospheric qualities that provided a 'measure of the affective appraisal of a space' (Stokkermans et al. 2018). This exemplifies methodologies in this field that investigate light's atmospherics by evaluating experience using surveys or that rate lighting conditions along scales of what feels 'safe' or 'pleasant' (Knight 2010), 'lively' or 'cosy' (Vogels 2008), and 'informal' or 'formal' (Li et al. 2019), or that use other, more detailed word pairings (Albertazzi et al. 2018).

Other studies venture outside the lab; for example, into retail environments (Custer et al. 2010), offices (van Duijnhoven et al. 2020) or outdoor urban environments. For example, in a study of lighting and perceptions of safety, van Rijswijk and Haans (2018) focused on questions of prospect, concealment and entrapment, a framework that recognised the importance of the lit surroundings and how people perceived and imagined what could happen in them. This study was set outside on a dim footpath at night, included the varying responses of different people and was one of several that purposefully included men and women of different ages (i.e., Johansson et al. 2010). It also included the element of movement, an aspect that often goes under researched, by assessing the accuracy of facial recognition as research participants moved along the footpath.

Although the lighting research discussed here investigates the experience of light, its methodologies necessarily predetermine the particular means for relating that experience because it is interested in how the experience of light relates directly to the design, colour, placement or other qualities of lighting technologies. In practice, this often means that research participants are asked to express their experiences in terms decided by the researchers; for example, on a scale for participants to rate or via a predetermined questionnaire. Crucially, this often does not allow participants to either qualify what these terms mean for them or, more importantly, nominate their own categories of feeling or sensory experience, categories that may draw in aspects of these settings that are not within the control of lighting designers.

Moreover, these studies, precisely because of their necessary orientation to the technologies of lighting design, only account for their effects in terms that are defined by and structured through the technologies. In other words, ways of experiencing light that do not directly address their technological aspects can fail to be fully accounted for by researchers or even communicated by participants because they are only asked to reflect on particular aspects of light and lighting. Aspects of lighting technology thus provide the sole terms for understanding illumination in this approach, which does not always probe the subtler experiential qualities of light that might draw in other senses, memories, impressions or anticipatory modes. Additionally, with a few exceptions, many of these studies are set in controlled laboratory conditions where the complex spatial, mobile, affective and sensorial contexts in which we usually experience light and lighting cannot be addressed.

This is understandable given a research approach that seeks to control variables in order to focus on technological or perceptual details, and is crucial in designing safe, effective and reliable lighting. However, this work can miss the vital social and cultural dimensions of the lit world, instead understanding human experience by means of predefined categories that direct attention at specific luminaires and their affordances, and how people perceive them against predetermined and limited criteria. While very particular (and valuable) questions about lighting design are asked and answered, such an approach often does not enable participants to express their experience of light in their own terms or move beyond the structures of the survey or questionnaire and provide new insights for technical development. Moreover, the uncertainty and contingency of everyday experience – which includes the environments in which lighting is most often encountered – is usually not able to be considered, even if these settings directly relate to the effects of lighting design. While I do not intend to unfairly critique this valuable research in lighting design, I argue that we now require approaches that also reach across disciplines, drawing lighting design and engineering studies together with in-depth and open-ended approaches that consider the perception and experience of light in everyday life, outside the lab. I discuss these and how they can augment existing research in the next section.

The lit world

Across the chapters of this book, the authors propose an approach to lighting that aims to account for the complexity of people's experiential worlds. To do so, several turn to the social sciences, where light is treated as inherent to how we apprehend and experience our surroundings, and how we situate ourselves both spatially and socially. In this sense, we dwell in and move through what Sarah Pink has termed a dynamic 'lit world' (Pink and Sumartojo 2018) that knots visual apprehension together with ways of feeling and understanding our surroundings and ourselves in them. The notion of the lit world recognises how, with a few exceptions, our experiential worlds are predominantly made sense of in and through light (or its absence). This is an approach concentrated on 'ways of looking and knowing *in* light and *through* light (Pink and Sumartojo 2018: 6; see also Ingold 2011), and chimes with Bille and Sørensen (2016: 160) who also understand light as 'not something you see, but something you see in'.

This understanding of the lit world is based in anthropology and human geography, where there are established concepts to reckon with the relationship between light and people's affective and sensory experience of their surroundings (Bille 2017; Ebbensgaard 2015, 2019; Slater et al. 2015; Edensor 2012, 2015). Here, light is central to how we make sense of the world and cannot be separated out from it. It does particular cultural work, such as signalling hospitality, social status or reverence (Bille 2014, 2017; Kumar 2015); it is socially important in bringing people together or marking out their difference (Edensor 2012); it is entangled with and signals different types of economic activity and governance regimes (Shaw 2014); it supports particular feelings, such as safety or vulnerability (Sumartojo and Pink 2017); and it marks out special places with particular aesthetic and symbolic importance (Edensor and Bille 2019; Sumartojo 2015). Cultural geographer Tim Edensor (2018), who contributes to this book a chapter on growing practices of 'dark design', identifies the many sensory and affective affordances of light, such as when people are drawn together in festive and convivial settings, or when harsh glare or unflattering tinctures cause anxiety or alarm (Edensor 2017). Indeed, anthropologist Tim Ingold (2011: 258) explains that light is foundational to human experience, 'a necessary precondition for the isolation of the perceiver as a subject with a "mind", and of the environment as a domain of objects to be perceived'. These are only a few of the many studies in anthropology and human geography that focus directly on light and lighting, their meanings, effects and potentialities.

Central to these understandings are the affective and sensory qualities of light that are core to how we perceive and respond to it. Almost all the chapters in this book are focused on urban settings, where the lit world is composed of different lights from multiple sources, such as buildings, cars, advertisements and signals that not only mix with each other but also bounce and reflect off surfaces, such as glass, stone and metal (Pink and Sumartojo 2018). This often messy and unsynchronised scene composes atmospheric effects configured from the different qualities of illumination, such as 'texture, accent, spatial transition, visual cues, security and perception of security, moods, cerebral temperature and drama', that generate various effects, including glow, glare and diffusion (see Figure 1.1) (Cochrane 2004: 12–13; Sumartojo et al. 2019).

An additional important consideration here is movement because the lit world often sees people or light sources mobile in relation to each other, and the visual and affective impacts appearing or being enlivened by this motion. This is alluded to in lighting design research, such as van Rijswijk and Haans' (2018) study in which participants were able to recognise well-known faces as they moved closer to them in dim outdoor night-time lighting conditions. In different ways, Sumartojo and Pink (2017) also discuss how feelings about lighting changed with proximity, as when people moved closer to a familiar light source, like a street sign that reminded them they were nearing home or moving car lights from the street outside refracted onto a bedroom window in a ghostly and beautiful display. Light's capacity to leak and mingle with other sources of light, both stationary and mobile, also capture the imagination and give rise to fleeting and coincidental displays, as with reflections of lights from office buildings in the dark, rippling water of a river at night (Sumartojo et al. 2019) (see Figure 1.1).

Accordingly, the concept of the lit world encompasses many of these factors: the wide variety of types, sources and conditions of illumination, including the way they mix with, interfere with or augment each other; the other things and people that they make visible; the interaction with the material world; the feelings and impressions that are associated with these apprehensions; the culturally specific ways of understanding and manipulating light and lighting; and the role of movement, be it of people, light sources or a combination of the two. By making human experience the chief focus of inquiry and by understanding people as dwelling and moving in a lit world in this way, new insights about lighting technology and design can become possible.

Such an orientation is in concert with methodologies that move outside of the lab; that treat lighting as embedded in a rich, multisensorial world; and that allow people to volunteer their own terms for making sense of and explaining their experiences of light and lighting that might exceed or even confound existing research frameworks. This, in turn, can form the basis for new design approaches that more fully account for these experiences and design into and with them. The chapters in this book provide a range of examples of recent projects that adopt this approach in different ways. While these were not always directly intended to make recommendations for lighting design, they form the basis for future research that can be targeted at the development of lighting design and technologies.

The shape of the book

The book begins with a direct argument for a new epistemology of lighting and then moves to two explorations of light and lighting in indoor hospital settings where light is part of how care and well-being are sensorially and atmospherically configured. From Chapter 5, we go outside, where the focus shifts to how light is related to feelings of safety, security and vulnerability in different urban night-time settings. The book finishes with two chapters that directly consider darkness as something that is designed both *for*, in the case of an urban high-rise building, and *with*, across a rich span of art and design-based projects. Together, the chapters collectively reinforce the value of research into lighting based on how it is experienced and how this connects to other complex feelings implicit in how people relate to each other in different settings, such as care and safety. It shows how designing with light (and dark) are intimately configured with these feelings and how the practice of lighting design could also be thought of as a social, spatial and experiential endeavour.

Ω The book starts with Nona Schulte-Römer's argument in Chapter 2 for a new epistemology for lighting, one that recognises the foundation of lighting standards in 'photometric experiments and statistical approaches', and that seeks to unsettle this by instead attending to how people use light for different purposes, how this is articulated in lighting practices, and how these result in subjective perceptions of light. In seeking to extend measurement-based understandings of lighting, she recognises the challenge of accounting for experiences that involve a 'co-production of sensuous light and sensing individuals, of affect and of atmospheres'. One way she does this is to discuss 'guerrilla lighting interventions' as a means to reach better understandings of how people experience light and to contribute new insights to lighting designer practitioners, planners and policymakers, an improvised approach also discussed in Chapter 6. This sets the scene for the exploratory and ethnographic approach to the experience of light and lighting that appears throughout the book.

Ω Chapter 3 takes us into a hospital setting and connects light and lighting directly to a sense of well-being. Advocating for a design anthropology approach, authors Sarah Pink, Melisa Duque, Shanti Sumartojo and Laurene Vaughan draw on an ethnographic project in the psychiatric wards of a public hospital to explore how engagements with light are valued, both in terms of everyday workplaces and the larger social and cultural values associated with light. To do so, they specifically address how natural daylight was articulated as supporting well-being in terms of professional values related to care and responsibilities for patients' and colleagues' well-being. This is a subtle account that invites the reader to reflect on how light is a matter of both good design and of 'the values and modes of innovating to create value that are integral to workplace cultures and practices'.

We stay in the hospital setting in Chapter 4 but move to a Danish birthing suite, focusing on chromatic lighting in maternity wards. Recognising the capacities of lighting to contribute to how people see and feel, in both a bodily and emotional sense, Stine Louring Nielsen emphasises how light can 'affect people's lives in the ways we act and feel, in inner bodily sensations and outer movements'. Starting with a conceptual foundation in the notion of atmospheres, she theorises how lighting can act on and with the body in different ways. She substantiates this with an account of the curation of lighting schemes by midwives during childbirth, including their manipulation of lighting colour schemes and fixtures for mood, to raise or lower the 'energy' in the room, and for specific clinical procedures. In a detailed ethnographic account, she shows how lighting is part of how midwives both attune to and help to adjust how the spaces of giving birth feel as part of their practice, purposefully using lighting to ongoingly shift the mood for women giving birth, their partners and others. As such, she argues that lighting is an element of design that works on the body in multiple ways, and that this needs to be better accounted for in regulatory approaches.

Chapter 5, by Hoa Yang, Jess Berry and Nicole Kalms, brings us into urban public space. Using a mixed-methods approach to understanding and measuring lighting, the chapter is an account of a unique project that started with crowd-sourced accounts of how women felt in different parts of the city after dark and then conducted detailed technical assessments of the lighting conditions in some of the places that were reported as feeling least safe. Led by lighting designer Hoa Yang, the chapter shows how different understandings of urban public space, represented through different data sets, can be fruitfully brought together. It argues that the omission in current standards of a consideration of the social and spatial contexts of lighting misses out 'the experiential impact that lighting can have on perceptions of safety in spaces'. Research that can better account for the contexts of lighting challenges existing best-practice urban design for the urban environment after dark and also supports moves towards more equal access to the night-time city by those who might otherwise feel unsafe or who might avoid certain places. The authors advocate for a practical approach that 'demonstrates how experience can be brought to a more conventional and technical language to inform design practitioners'.

We stay with feelings of safety and lighting in the night-time city in Chapter 6 with a project related to the one described in Chapter 5, in an account of a lighting design workshop conducted in Melbourne's city centre. Using an ethnographic methodology related to that of Chapter 2, this chapter, by Shanti Sumartojo, treats light as directly constituent of how places feel to people, exploring this through an after-dark workshop that asked people to design their own lighting schemes for an inner-city laneway. The chapter is based on the premise that how safe we feel in the urban built environment is shaped by continually emerging factors that lighting design contributes to, but cannot comprehensively determine. Based on the participatory co-design workshop it describes, it argues that light and lighting are made sense of *emergently*, which unsettles approaches in lighting design that seeks to plan what light will look like and what effects it will have. The chapter reinforces the conclusions of the book that lighting design should be driven not only by practical concerns and regulatory stipulations, but also by ideas of care, safety and belonging that can only be understood in the specific contexts of the lit world.

In Chapter 7, Mikkel Bille and Olivia Norma Jørgensen take us back to Denmark, this time to a city square in Copenhagen called Blågårds Plads. As with many of the other contributions, they consider how lighting is part of the complex social and cultural aspects of belonging in the square, particularly in terms of security and neighbourhood identity, and analyse attempts to reflect and improve this through both practical and artistic lighting. The authors enrich the atmospheric conceptual framework proposed in Chapter 4, using the notion of urban atmospheres to explore different effects and affects of light in this complex site. They discuss accounts from residents for whom light is central to how they make sense of their lives and activities in the square and alongside the other people there, including those who might be engaged in criminal activity. They show how the square must be understood as 'a multiplicity of felt spaces that unfold in different ways at different times', a dynamic and multivalent rendering that complicates urban planning or lighting design for an apparently fixed urban built environment. As in Chapter 5, they make an argument for engaging with how lighting *feels* to reach the best outcomes for how to design with and for it, and to take account of the complexity of an urban site.

If lighting in a city square can speak to the experiences of where people live and work, then the ways they move around the city are also a vital part of everyday experiences of lighting. Chapter 8 provides a case study of a tram stop in the Danish city of Aarhus, where the research team was able to directly manipulate the lighting and investigate the effects of different light levels for public transport users. Mette Hvass, Karen Waltorp and Ellen Kathrine Hansen consider the dimming of light, in a focus on the possibility of darker lightscapes, a topic that the following chapters also take up. They contend that lower light levels have the capacity to contribute to a more relaxed atmosphere at their research site and to help visually connect the tram waiting area and the surrounding urban context. As in previous chapters in the book, feelings of safety and comfort emerge as central to how people sensorially experience and understand light. They echo arguments found in Chapters 5 and 6 that dimmed lighting that decreases visual contrast with the surroundings and that is better balanced with the brightness of proximate light sources can improve not only how spaces feel but also how people interact in them.

Chapter 9 goes vertical and brings a multiscalar exploration of a high-rise residential tower into its discussion. Casper Laing Ebbensgaard investigates how such towers are designed specifically for night-time conditions and draws together the ambitions and visions of architects and designers – who are mostly interested in the visual impact of the building on the surrounding urban environment – with the everyday experience of living in a building where cutting-edge lighting technologies are incorporated throughout the building. The chapter asks how the design of residential towers for their appreciation by others at night and their inhabitation by residents can advance an understanding of them as contributing to the formation of an urban public night as commons. To do so, Ebbensgaard moves between accounts from designers and from residents, thereby locating the building within a shared urban environment where light works to connect people to each other and to their surroundings. This argument relates to those in Chapter 7 about the multivalence of the lit world because it shows the entanglement of light in sociality in multiple ways.

Tim Edensor closes the book in Chapter 10 with a detailed discussion of designing with darkness. He argues that we now live in 'a vastly over-illuminated world dominated by notions of safety, standardisation, unequally distributed lighting of poor quality and light clutter', and calls for more attention to the 'ways in which we can design with shades of darkness'. Thus, instead of directing his gaze at light, its regulation, control or design, he instead peers into the gloom, pointing out how artists, architects and designers have always worked with darkness as much as with light and illumination. Recognising the importance of darkness in any consideration of the lit world, he focuses on the manipulation of shadow and shade, foregrounding how light and dark are inextricably relational qualities that suffuse our everyday lives. He also shows how these cannot be separated from how we feel about and experience our surroundings, aligning with arguments in other chapters about the importance of dimmer lighting settings for shared public spaces that could thereby feel safer and more inclusive, consume less energy, and support the flourishing of other species with which we share our environments.

Conclusions

The parallel developments of work on light and lighting in the fields of human geography, anthropology, design and lighting engineering suggest an interdisciplinary way forward for new ways to understand the relationship between lighting and how it feels. This book seeks to work into and through different disciplinary ways of understanding how people experience their lit environments so that the complexity of human perception of light might be better directed to the technical development of luminaries, their design, placement and interaction with specific settings. In other words, the book argues that accounting for experience in a more expansive way can improve lighting design. This is particularly the case where designers aim to support a sense of safety, comfort or conviviality, feelings that lighting design has been shown to demonstrably impact. Accordingly, in this book, the authors contend that a refigured approach can offer valuable new insights that cannot be reached by other methodologies in isolation. By working in interdisciplinary combination, new ways of both conceptualising and designing lighting can emerge, be implemented and contribute to how we make sense of and feel in and about our own lit worlds.

References

Albertazzi, L, Canal, L, Chisté, P, Micciolo, R and Zavagni, D (2018) Sensual Light? Subjective Dimensions of Ambient Illumination. *Perception* 47(9): 909–926.

Bille, M (2014) Lighting Up Cosy Atmospheres in Denmark. *Emotion, Space and Society.* http://dx.doi.org/10.1016/j.emospa.2013.12.008

Bille, M (2017) Ecstatic Things: The Power of Light in Shaping Bedouin Homes. *Home Cultures* 14(1): 25–49.

Bille, M and Sørensen, T (2016) A Sense of Place. In M Bille (Ed.), *Elements of Architecture: Assembling Archaeology, Atmosphere and the Performance of Building Spaces.* London: Routledge, pp. 159–212.

Cochrane, A (2004) Cities of Light Placemaking in the 24-hour City. *Urban Design Quarterly* 89: 12–14.

Custers, PJM, de Kort, YAW, IJsselsteijn, WA and de Kruiff, ME (2010) Lighting in Retail Environments: Atmosphere Perception in the Real World. *Lighting Research and Technology* 42: 331–343.

Ebbensgaard, C (2015) Illuminights: A Sensory Study of Illuminated Urban Environments in Copenhagen. *Space and Culture* 18(2): 112–131.

Ebbensgaard, C (2019) Making Sense of Diodes and Sodium: Vision, Visuality and the Everyday Experience of Infrastructural Change. *Geoforum* 103: 95–104.

Edensor, T (2012) Illuminated Atmospheres: Anticipating and Reproducing the Flow of Affective Experience in Blackpool. *Environment and Planning D: Society and Space* 30: 1103–1122.

Edensor, T (2015) Light Design and Atmosphere. *Visual Communication* 14(3): 331–350.

Edensor, T (2017) *From Light to Dark: Daylight, Illumination and Gloom.* Minneapolis: University of Minnesota Press.

Edensor, T (2018) Moonraking: Making Things, Place and Event. In L Price and H Hawkins (Eds.), *Geographies of Making/Making Geographies: Embodiment, Matter and Practice.* London: Routledge.

Edensor, T and Bille, M (2019) 'Always Like Never Before': Learning from the *lumitopia* of Tivoli Gardens. *Social & Cultural Geography* 20(7): 938–959.

Gray, C, Kou, Y, Battles, B, Hoggatt, J and Toombs, A (2018) *The Dark (Patterns) Side of UX Design.* CHI 2018, April 21–26, 2018, Montreal, QC, Canada. https://doi.org/10.1145/3173574.3174108

Ingold, T (2011) *The Perception of the Environment: Essays on Livelihood, Dwelling and Skill.* London: Routledge.

Jensen, R, Strengers, Y, Raptis, D, Nicholls, L, Kjeldskov, J and Skov, M (2018) *Exploring Hygge as a Desirable Design Vision for the Sustainable Smart Home.* DIS'18, 9–13 June 2018, Hong Kong. https://doi.org/10.1145/3196709.3196804

Johansson, M, Rosén, M and Küller, R (2010) Individual Factors Influencing the Assessment of the Outdoor Lighting of an Urban Footpath. *Lighting Research and Technology* 43: 31–43.

Knight, C (2010) Field Surveys of the Effect of Lamp Spectrum on the Perception of Safety and Comfort at Night. *Lighting Research and Technology* 42: 313–329.

Kuijsters, A, Redi, J, de Ruyter, B, Seuntiëns, P and Heynderickx, I (2015) Affective Ambiences Created with Lighting for Older People. *Lighting Research and Technology* 47(7): 859–875.

Kumar, A (2015) Cultures of Light. *Geoforum* 65: 59–68.

Li, B, Zhai, QY, Hutchings, JB, Luo, MR and Ying, FT (2019) Atmosphere Perception of Dynamic LED Lighting Over Different Hue Ranges. *Lighting Research and Technology* 51: 682–703.

Pink, S and Sumartojo, S (2018) The Lit World: Living with Everyday Urban Automation. *Social & Cultural Geography* 19(7): 833–852.

Shaw, R (2014) Streetlighting in England and Wales: New Technologies and Uncertainty in the Assemblage of Streetlighting Infrastructure. *Environment and Planning A* 46: 2228–2242.

Simonsen, J (2018) User Experience. In K Norman and J Kirakowski (Eds.), *The Wiley Handbook of Human Computer Interaction*, Vol. 1. Hoboken, NJ: Wiley, pp. 191–206.

Slater, D, Sloane, M and Entwistle, J (2015) *Configuring Light: Staging the Social*. http://www.configuringlight.org/ Accessed 18 August 2021.

Stokkermans, M, Vogels, I, de Kort, Y and Heynderickx, I (2017) Relation between the Perceived Atmosphere of a Lit Environment and Perceptual Attributes of Light. *Light Research and Technology* 50(8): 1164–1178.

Stokkermans, M, Vogels, I, de Kort, Y and Heynderickx, I (2018) A Comparison of Methodologies to Investigate the Influence of Light on the Atmosphere of a Space. *LEUKOS* 14(3): 167–191.

Sumartojo, S (2015) On Atmosphere and Darkness at Australia's Anzac Day Dawn Service. *Visual Communication* 14(3): 267–288.

Sumartojo, S and Pink, S (2017) Moving through the Lit World: The Emergent Experience of Urban Paths. *Space and Culture* 21(4): 358–374.

Sumartojo, S, Edensor, T, and Pink, S (2019) Atmospheres in Urban Light. *Ambiances* 5: 1–20. https://journals.openedition.org/ambiances/2586?lang=en.

van Duijnhoven, J, Aarts, MPJ and Kort, HSM (2020) Personal Lighting Conditions of Office Workers: An Exploratory Field Study. *Lighting Research and Technology* 53(4): 285–310.

van Rijswijk, L and Haans, A (2018) Illuminating for Safety: Investigating the Role of Lighting Appraisals on the Perception of Safety in the Urban Environment. *Environment and Behavior* 50(8): 889–912.

Vogels, I (2008) Atmosphere Metrics: Development of a Tool to Quantify Experienced Atmosphere. In JMDM Westerink, M Ouwerkerk, TJM Overbeek, WF Pasveer and B de Ruyter (Eds.), *Probing Experience*. Dordrecht: Springer, pp. 25–41.

Chapter 2

ILLUMINATING EXPERIENCES

Lighting design as an epistemic approach

Nona Schulte-Römer

DOI: 10.4324/9781003182610-2

Nona Schulte-Römer

Introduction: exploring an 'epistemic wasteland'

Today, the majority of the world population lives in cities, which are often illuminated 24 hours a day, at least in industrialised parts of the world (Melbin 1987; Shaw 2015). Yet while so many people are exposed to artificial light at night, we still know fairly little about how illumination is experienced on an everyday basis in specific public spaces. Although lighting shapes our everyday lives and experiences in profound and culturally specific ways (Bille and Sørensen 2007), it tends to escape our attention and, hence, our influence. While *making light* has become an increasingly functionally differentiated expert task and 'sunk into the background' (cf. Star and Ruhleder 1996), *living with light* in cities is taken for granted. People in public spaces are rarely aware of the lighting, let alone involved in the light planning of their everyday environments. This lack of engagement in urban illumination is further intensified through a division of labour and diversification of lighting functions. Public or public-private services provide street lighting; numerous private actors, including commercial shop owners, the advertising industry or building managers, illuminate the nocturnal city with a myriad of lights.

While functional lighting has become the domain of technicians and engineers, architects, designers and artists are supposed to take care of atmospheric, decorative or spectacular lighting schemes. The geographer Hasse (2007) thereby observes a biased focus on mechanistic, administrative and legal processes that coincides with a deficit in 'aesthetic knowledge'. This deficit goes beyond disciplinary differences between expert groups or the notorious knowledge gaps between experts and laypersons. Instead, Hasse describes an 'epistemic wasteland' that concerns the subjective experience of light in space as an object of aesthetic planning and design (Hasse 2007: 71–72). While we *know* the costs, technological requirements or planning regulations of lighting projects and can express them in figures, numbers or paragraphs, it is difficult to tell what a person feels when s/he experiences an illuminated space. In the absence of objectified evidence and aesthetic knowledge, lighting designers rely on their tacit professional knowledge and reference projects in order to communicate or justify their aesthetic choices or the cost of their designs.

Figure 2.1
Berlin's Alexanderplatz
at night: an orchestra
of public and private
lights.
*Photo: Nona
Schulte-Römer.*

In this chapter, I outline how lighting designers balance this deficit in innovative and participatory ways. First, to this end, I explore this 'epistemic wasteland' in historical perspective as a process of multiple fragmentations in lighting. Focusing on contemporary lighting practices, I then argue that professional lighting design can offer the means and methods to explore and restore subjective experiences in light planning practice and processes. More precisely, I propose that lighting designers' professional *focus on specific lighting situations* draws attention to the experience of multiple light sources in space, irrespective of whether the sources are 'functional' or 'decorative' lighting. Second, I suggest that the 'epistemic culture' (Knorr-Cetina 1999) of lighting design can integrate light-related knowledge from different disciplines, procedural expertise in planning as well as a *professional reflection on subjective experiences* of light in space. Third, I argue that this epistemic wasteland cannot be explored in the laboratory or in one big academic mapping exercise but requires a constant and inclusive exchange about subjective experiences. In this regard, I consider *lighting design methods, like site visits, night walks and experimental installations*, as a productive way forward. Finally, I show how an exploration of the 'epistemic wasteland' of urban illuminations is relevant not only for geography and urban planning but also for current trends in urban lighting. In a situation where the lighting field is transformed by the 'LED revolution', climate change mitigation policies and discourses on light pollution, reconsidering our experiences of and needs for light and darkness seems a prerequisite for developing sustainable solutions in lighting.

The argument I develop in this chapter builds on my social-scientific research in the lighting field. This data collection started with an expert workshop[1] on urban light planning (Schulte-Römer 2010) and continued during my research on the introduction of LED street lighting (Schulte-Römer 2015) and transdisciplinary research on light pollution (Schulte-Römer et al. 2018). In this period, numerous lighting design events and conferences offered great occasions for ethnographic research.

Urban lighting fragmented – a historical review

In order to understand how and why the subjective experience of light remains an epistemic wasteland, it is worthwhile taking a historical view of the fragmentation of lighting purposes, disciplines and perceptions. In Western society, the conquest of the urban night really started in the 17th century when European absolutist rulers invented night-time entertainment – at the time, an extremely luxurious alternative to daytime festivities as candles and oil lamps were very expensive and only the upper classes could afford to illuminate the night and sleep during daytime working hours (Brox 2010; Koslofsky 2011). During industrialisation, the implementation of gas lighting infrastructures and, later, electricity (Schivelbusch 1988; Tomory 2009) made artificial lighting widely available and affordable. In this process, however, urban lighting purposes, practices and perceptions of light also became increasingly fragmented.

First, industrialisation promoted a *fragmentation of lighting purposes* into what is often described as 'decorative' and 'functional' lighting.[2] Functional area and road lighting became increasingly standardised and tied to photometric benchmarks, like brightness and uniformity levels, to guarantee good visibility in public spaces (Otter 2008). The co-evolution of electrification and motorised traffic in the early 20th century spurred this traffic and safety-oriented function-alisation of public lighting (Isenstadt 2018; Jakle 2001). Meanwhile, decorative festive, architectural or spectacular illuminations became more sophisticated and attracted ever-larger audiences (Binder 1999; Nye 2018; Barnaby 2009). Architectural lighting, commercial illumina-tion and shining screens became a decisive feature of the 'modern' and, later, the 'post-industrial' and 'aestheticized' city (McQuire 2008; Reckwitz 2009; Schulte-Römer 2011; Edensor and Bille 2017; Eden-sor 2012; Dannemann 2017). In the 1990s, municipalities increasingly rediscovered architectural and festive illuminations as a powerful tool for city beautification (Gonzalez 2010).

Second, the differentiation of lighting purposes coincided with a formation of lighting institutions that paved the way for frag-mented lighting practices, light-related knowledges and professional groups. In the early 20th century, light engineering societies formed in industrialised countries around expert tasks and challenges, like the photometric measurement of light in space (Johnston 1994). Around the same time, avant-garde artists, architects and designers, like László Moholy-Nagy, Bruno Taut, Le Corbusier, Ludwig Mies van der Rohe and others, discovered natural and artificial light as an impor-tant building material and tool for creating atmospheres and spaces (Hoormann 2003; Isenstadt 2018; Hirdina 1997). With this speciali-sation, electrical engineers and lighting designers developed distinct 'epistemic cultures' with their own ways of producing light-related knowledge and light (Knorr-Cetina 1999). The subjective experience of light in space became the domain of artistic or architectural lighting practice, while large-scale lighting infrastructures were planned on the basis of laboratory experiments and reflected the visual perfor-mance of the human eye under standardised conditions.

Third, the professionalisation and delegation of light planning and lighting design to experts also created a *fragmented perception of urban lighting*. While lighting experts developed a technologically enhanced and self-reflexive 'professional vision' (Goodwin 1994; Otter 2008), ordinary people or 'light consumers' increasingly turned a blind eye to the 'invisible infrastructure' (Latour and Hermant 2006 [1998]; Pinch 2010) that illuminated their world at night. This has not only contributed to widespread 'inattentional blindness' (Zerubavel 2015) towards lighting matters. It has also created a language bar-rier between lighting professionals, who can eloquently express their views on light, and laypersons, who usually lack the vocabulary to articulate what they see and feel; and when they complain, infrastructural refurbishments are often already on their way or completed (Green et al. 2015; Ebbensgaard 2019; Besecke and Hänsch 2015; Schulte-Römer 2015). As a result of lacking exchange between those who make light and those who live with it, urban illuminations and city lights have, in some places, become matters of public concern and caused controversies (Hasenöhrl 2015; Challéat et al. 2015; Morgan-Taylor 2015; Meier 2019).

Lighting design as a boundary-crossing epistemic culture

In the context of this multiplicity of purposes, forms of knowledge and perceptions of lighting, lighting design as a practice and discipline offers the possibility of contributing to a less fragmented understanding of city lights as a sociocultural phenomenon. This boundary-crossing also implies what sociologists call 'boundary work'. The concept describes how actors, including professionals and professional associations and organisations, shape identities and gain authority by performing their particular practice in distinct and reflexive ways and by making public claims that distinguish them from others (Gieryn 1983; Lamont and Molnar 2002).

Boundary work in lighting design, thus, includes all activities that define, distinguish and institutionalise lighting design as a profession and professional practice. Lighting design is a rather young and hybrid discipline with roots in architecture, light engineering, media design, art and stage lighting (Hansen and Mullins 2014). Professional programs for lighting designers only emerged in the 1960s. Since then, international associations have stabilised and grown (IALD) or vanished (ELDA and PLDA) and facilitated internal discussions, exchange and professional self-reflection. During lighting design conferences, the standardisation of educational formats and discussions on the role of lighting design within a more engineering-oriented lighting industry are reoccurring themes. What is remarkable about lighting designers' boundary work is that their claims, professional practice and approaches not only draw a line between conventional, non-designerly lighting practices. Instead, they appear suited to 'defragmenting' perspectives on urban lighting and exploring subjective experiences of light as aesthetic phenomena by 1) seeing light in space; 2) integrating and mediating disciplinary, procedural and aesthetic knowledge; and 3) engaging with laypersons and their experience of light in space.

Seeing functional lighting aesthetically

The fragmentation of lighting purposes into visible decorative, architectural and commercial illuminations on the one side and invisible functional area lighting on the other coincides with a fragmentation of empirical methods of light-related evidence production. Decorative lighting acknowledges 'subjective experiences' in a qualitative, often non-formalised way. Functional lighting, in contrast, is planned and produced on the basis of technical standards and performance measures. As historians have wonderfully illustrated (Otter 2008; Johnston 1994; Jakle 2001), these standards were developed on the basis of photometric experiments with the explicit intention to objectify subjective light experiences. Moreover, the standardisation of measurement procedures objectified the visual perception of those who planned and maintained public lighting infrastructures. The evidence-based standardisation and professionalisation of light planners' vision created what Lash (2018) describes as 'objective experience'. Today, standard light planning no longer depends on lighting professionals' subjective sensory perceptions, let alone their aesthetic experience of an urban space. Instead, they plan lighting fixtures on their desks by calculating standard values for brightness levels, uniformity and the efficiency of lighting systems (Jakle 2001).

During my field research (2011–2012), I had the chance to accompany several lighting professionals – engineers and lighting designers – during their site visits to urban LED installation sites (Schulte-Römer 2015). While they inspected the sites and light quality, I asked questions to understand what they saw and how they evaluated the newly installed innovative technology. The difference between engineers' ways of seeing and lighting designers' approaches was obvious. In a nutshell, the lighting engineers directed my attention to measurable characteristics. In their eyes, an LED installation passed the test if the luminaires did not produce glare, if the lux level on the street was reasonably high and, very importantly, if the uniformity of the illumination was good (i.e., if there were no dark spots between the light poles). Whereas the lighting engineers could estimate fairly well with their bare eyes how brightly (in lux) a street was lit, I could not even come close to this kind of 'objectified experience' due to my lack of training and experience. I was even unable to determine the brightness of the street in an 'objectified' way when a municipal engineer handed me a lux metre. The lux levels on the display differed considerably depending on how I held the metre, and I realised that producing light-technologically meaningful, reproducible values required a great deal of procedural knowledge.

During site visits with lighting designers in contrast, I found that the focus was more on an assessment of the overall urban situation and experience of night-time spaces, and not just on the technology performance. While designers also touched issues like glare, uniformity and brightness levels, they often actively invited subjective experiences of public spaces even when functional lighting was concerned. One night in Lyon, I attended a site visit on a local university campus guided by a renowned local lighting designer. While we walked through the space, he critically reflected on why certain luminaires were installed in specific places and not in others (mostly for cost reasons), what their functions were (mostly security aspects) and how the users of that space interacted with the light (high potential for vandalism). After the tour, I took the following notes:

> It is surprising how much I have overlooked, although I pass through this space almost every night. The lamp masts are very high. Therefore, the light spills all over the place up to the third floor of the adjacent office buildings. According to [lighting designer], the lamp posts should only be half as high for this type of street. I wonder if this is why I have the impression that this is a sad place – at day and at night.

A few months later, the German light technological society held its annual conference Licht 2012 in Berlin. One night, the program offered a night walk in the city centre. Our heterogeneous group of lighting professionals and academics was guided by a well-established Berlin lighting designer who had also illuminated some of the buildings and places we passed. After we had inspected the illuminations at Brandenburger Tor and along the grand boulevard Unter den Linden, we finally stopped at Bebelplatz, a square that is surrounded by Prussian architectural monuments, such as the Berlin National Opera, the St. Hedwig's Cathedral and the main buildings of Humboldt University. The lighting designer who guided us pointed out that the illuminated wide boulevard and the functional lighting on the square created a visual barrier that made it difficult to experience the architectural ensemble. He recommended dissolving the barrier by only lighting the facades – as it was done on Place de Terreaux in Lyon – and lowering the light levels of the functional illumination on the square. To his disappointment, Berlin's administration had decided to stick with the standard and maintain the recommended lux levels for pedestrian areas, ignoring that the architectural lighting provided additional light sources that added to the lux levels on the pavement. This critique of standard lighting practice was followed by controversial discussion among the participants of the tour. While in fact, the industry standard EN 13201 is not legally binding, two municipal engineers insisted that the Berlin administration was obliged to implement the European road lighting standard. The episode shows that the standard is not very well understood by municipalities, has gained a law-like effect on municipal lighting practice (cf. Chain 2010) and can prevent a holistic assessment of lighting needs and urban situations.

To sum up, the two examples of night walks illustrate how lighting designers redraw the boundaries between decorative and functional lighting by seeing street lights through an aesthetic lens, with an emphasis on the subjective experience spatial situations. This way of seeing often clashes with objective ways of evaluating lighting as explicated in lighting standards, like the EN 13201 for road lighting, which was developed on the basis of light-technological knowledge and photometric experiments that focus on visual performance – or how well the average human eye can see – as their key criterion. The lighting designers I have met are aware of lighting standards and their functional value. But they look beyond standardised computer calculations and focus on the experience of light in real-world situations. This is in line with the findings of Casper Laing Ebbensgaard (2020: 1957), who argues that 'designers can play an important role in challenging how standards and regulations are measured, defined and maintained'.

Lighting designers' professional vision and approaches were also discussed during our expert workshop (Schulte-Römer 2010).

A lighting designer explained that she spends a lot of time observing city lights. As a result of training her perception in this way, she now routinely recognises the light levels rather than the material objects in a space. Regarding standards, one lighting designer explained that 'one has to hack the standard' in order to produce good lighting design. He further explained that such reinterpretations presuppose an in-depth knowledge of the standard and the 'objective' photometric principles inscribed in it. This also presupposes a trained professional vision that brings aesthetic experiences to light. In this sense, lighting designers' professional way of seeing light in space (Goodwin 1994) can be considered as a boundary-crossing practice that cherishes the aesthetic appeal of light in space – even if it is 'just' street light.

Integrating disciplinary, procedural and aesthetic knowledge
The lighting design profession seems particularly well suited to crossing disciplinary boundaries and integrating knowledge and practices from different light-related fields. Lighting design has been a transdisciplinary profession from the start, incorporating and bridging the boundaries between different 'epistemic cultures' (Knorr-Cetina 1999). Designing light requires knowledge in areas like architecture, urban planning, physiology and engineering as well as entrepreneurial and design skills. I have met architectural lighting designers who proudly present their professional backgrounds in electrical engineering. Educational programs reflect the transdisciplinary knowledge base. For instance, the international Lighting Design Master Program at Wismar University of Applied Sciences encourages students 'to investigate artistic, physical and psychological aspects of lighting, using creative and autonomous methods'. As outlined on the website of the program, the students also learn 'market analysis tools and the strategic management approach' so that they can develop a 'vision of design as part of the customer value chain' (Wings professional studies 2021).

At the same time, lighting design is still a profession in the making. While influential light engineering societies, like the IES in the USA, the ILP in the UK, CIE in France and LiTG in Germany, have recently celebrated their 100th anniversaries, lighting design associations are much younger. The International Lighting Design Association (IALD) was only founded in 1969 in Chicago, USA. As stated on the website, IALD 'strives to set the global standard for lighting design excellence by promoting the advancement and recognition of professional lighting designers'. Lighting design events, like the IALD conference Enlighten Americas or the PLDC (Professional Lighting Design Convention), give lighting designers the opportunity to present their work, discuss what defines good practice and quality in lighting design, and exchange views about the potentials and challenges of technological developments, like LED and adaptive lighting, and other new topics and knowledge.

An emerging knowledge area of increasing importance is the transdisciplinary research field that investigates the negative side effects of artificial light at night (ALAN) on ecosystems and human health (Hölker et al. 2010; Pérez Vega et al. 2021; Zielinska-Dabkowska and Rhode 2017). This research has spurred public discourse on light pollution and challenges lighting designers to position themselves. During the 2010 expert workshop, one observed that good lighting design respects the value of darkness, but dark sky movements had become an important player and sometimes made his work more difficult: 'We used to be magicians, now we are guilty, the bad guys'. This seems to be changing. During our more recent survey on light pollution (Schulte-Römer et al. 2019), more than one third of our self-selected respondents were lighting designers from all parts of the world. Many of them actively tackle light pollution (Schulte-Römer et al. 2018). When asked what they recommended in order to avoid light pollution, they highlighted the need for knowledge and information exchange. In particular, they considered education of lighting professionals and clients, and integrated light planning and best-practice examples more viable than better technology or concrete standards or legislation. They also repeatedly pointed out that obtrusive light and light pollution could be reduced if lighting was planned by professional designers. 'Hire a lighting designer!' was a feedback we received in various variations. One survey respondent argued that 'architects and architectural lighting designers *are* environmental designers by definition – they have the professional and moral obligation to safeguard the environment'.

Lighting designers' view on the 'loss of the night' thereby often adds an aesthetic touch to the environmental discourse. While the environmental perspective focuses on the adverse effects of artificial light at night on flora, fauna and humans, and the risk of bad health and biodiversity loss, lighting designers' narratives also highlight the loss of a subjective experience of night skies and darkness. For instance, Paulina Villalobos, a Chilean lighting designer and founder of the project *Noche Zero*, explains her engagement for dark skies as follows:

> It is a thing of very mixed emotions. I grew up watching the stars. They were part of my daily environment. I grew up in a small town in the Atacama Desert, which is a very nice place to see the stars. So, when I moved away to study lighting design in Europe and Japan, I was missing something.

> (Schulte-Römer et al. 2018: 47)

French lighting designer Roger Narboni describes such a dystopia of pervasive brightness in a short science fiction novel *La Nuit Disparue*: In the year 2055, a European city is brightly illuminated around the clock. Darkness, the night or shadows no longer exists, not even as a concept. 'Without shadow the perception of the environment is distorted. Bodies and buildings seem to float in space without any possibility of understanding their true position and ground. . . . Without the shadow, without darkness there are no contrasts and all lights become uninteresting' (Narboni 2009: 8). Lighting design has become a lost cause, but it is also a source of hope. When a small group of artist rebels starts an 'innocent and relaxed neighbourhood festival of lights', which performs the *absence* of light as its key theme, the event develops into a major attraction and international movement. 'The sons, daughters or the grandchildren of former lighting designers' join the movement that stands 'for the return of darkness' and rebellion 'against the official lighting' (Narboni 2009: 10).

Narboni's novel as well as his concept of 'dark infrastructure' (Narboni 2021; Edensor 2017) offer a perfect example of how lighting designers integrate scientific findings into their design practice and narratives. The issue of light pollution is also a good example for how lighting designers perform a mediating position between opposing lighting needs, between positive and negative, and between the visual and non-visual effects of light at night. They can play this role not only because they know light and darkness from different disciplinary perspectives but also because they sell designs not products. In order to achieve good results, they move between light manufacturers and clients,

Figure 2.2
Science fiction:
lighting designer
Roger Narboni
describes the
dystopia of a world
without shadow in his
graphic novel *La Nuit
Disparue*.
Image: Roger Narboni.

act as knowledge brokers and engage in technology co-development. Accordingly, one of the lighting designers in our workshop described his role in the context of urban light planning as close to the job of a moderator who facilitates new connections within the city by starting the communication between different departments and stakeholders.

To sum up, the integration of knowledges (plural) from different disciplines and the focus on light in space seems characteristic for the epistemic culture of lighting designers. Yet while incorporating photometric and ecological knowledge, the key reference for evidence production in lighting design remains the specific project with its specific lighting situation and constellation of actors. This is also where sustainability claims are tested and negotiated. A lighting design is not automatically sustainable because it supports local commerce or, at the other extreme, reinstalls natural darkness for the sake of wildlife. Instead, the art of 'good' lighting lies in finding the right equilibrium not only between light and darkness but also between different and often contradictory lighting needs. The lighting design profession can contribute to this process by planning and moderating projects on the basis of light-related knowledge from various disciplines and by highlighting the *specific lighting situation* as the site where all evidence and claims to light and darkness need to be negotiated and mediated. The engagement of citizens as those who eventually live under and with new lighting designs can contribute to sustainable planning and lighting.

Engaging citizens in their familiar world

Although urban spaces are eventually illuminated for people, they are rarely actively involved in lighting projects and design decisions. During our workshop, a lighting designer reported that it is not easy to convince cities to invest money to illuminate residential areas rather than churches or town halls. Moreover, energy and cost savings are often more important than the question of how people experience their neighbourhood after dark – especially when it comes to functional lighting projects. However, while technical standards and photometric insights can offer guidance on how to achieve visual comfort and how to avoid glare, they tell us little about how humans sense light with their body and make sense of it based on their knowledge, experiences and culture. In Berlin, I met street light amateurs who told me that the sound of streaming gas contributed to the special atmosphere in West Berlin's gas-lit streets. In Lyon, a resident told me that the new LEDs in front of his house sparkled like Christmas lights. When I ask 'laypersons' what light colour they prefer, they often opt for warm-white or yellowish light, even if the light source is a sodium vapour lamp with a bad colour rendering (cf. Ebbensgaard 2019).

This approval of sodium lighting contradicts the views of lighting professionals and, thus, nicely illustrates the gap between lay and expert perceptions. Lighting professionals usually consider white full-spectrum lighting as the better choice for public spaces as it has a better colour rendering and allows people to see better at lower light levels. Yet, when I once confronted a lighting designer with what I suspected was a disregard of people's popular opinion (cf. Edensor and Millington 2013), he assured me that in his work with local communities, people would always opt for the white full-spectrum light when he let them directly compare between a sodium and full-spectrum illumination. In other words, laypersons might reconsider their habitual opinion about lighting and explore better options when they are offered the opportunity to engage with lighting and when they are asked about their perceptions and opinions.

Lighting designers have developed numerous formats and projects for engaging citizens in such explorations. Some formats are mini-experiments in transforming familiar worlds with light. What they have in common is that they let people experience light in real-world situations. During my research, I had the chance to witness several participatory design interventions, including the presentations and interventions of the first Social Light Movement Workshop in Liège, Belgium (Social Light Movement 2011), as part of the LUCI 'City under the Microscope' conference. The Social Light Movement is an initiative that promotes light as 'a right, not a privilege' and 'People before places' (www.sociallightmovement.com). In the course of the workshop design, students and the lighting designers[3] created concepts for specific urban locations in Sclessin, a suburb of Liège in Belgium, where the city saw urban planning potential. As outlined in the workshop documentation, the participants 'undertook mock-ups, liaised with residents and finally presented their proposals to the City of Liège and to the residents'. They also 'spent time walking, talking and playing football with kids and teenagers in Sclessin to understand how their lives are shaped by the area'.

Figure 2.3 a
Urban residents are often used to yellow sodium vapour lamps (image a). Lighting designers promote light with a better colour rendering for urban public spaces (image b).
Photos: Nona Schulte-Römer.

Figure 2.3 b

The main objectives of the workshop were to teach participants ways and tools to engage with the community, to design urban lighting and 'something more' in difficult and neglected areas, forcing participants to go out, to talk, to design for the people and not for themselves and to help convince cities and investors that there is another way of working! (sociallightmovement.com)

During the conference, urban light planner Isabelle Corten further outlined how light walks and site visits with citizens promote a better understanding of how the people who will be affected by urban and lighting design perceive these environments and how they use them after dark. Night walks can thereby help mediate between the lighting designers' professional vision and the lay participants' personal perspectives. While 'experts of the everyday' can bring in their deeply embedded and situated knowledge about the practical uses and meanings of a visited place, experts in lighting might open their eyes to aspects they have never recognised before.

On the closing night, local citizens and conference participants had the chance to take part in a Guerrilla Lighting intervention guided by the lighting designers of the Social Light Movement. During that intervention, the group used portable light sources, like spotlights, torches and battery-powered LEDs, to transform mundane and rather uncanny urban spaces for a few minutes, and thus create a completely different atmosphere. While the intervention itself was ephemeral, the participants' experience was lasting, and the Guerrilla Lighting Team made sure that the installation was photographed. The purpose was to raise people's awareness of the possibilities and transformative power of light in space (Light Collective 2008).

To sum up, engaging citizens in lighting projects is not easy, as it takes time and money to create suitable formats, and analytical translation to make the outcomes directly applicable to lighting design practice. Nevertheless, lighting designers have found formats and funding to promote participatory design as good professional practice. Site visits and light walks are thereby an important means of helping lay participants to see light in the first place, to acknowledge its performative power in their familiar environments and to find words to express their experiences. It is also a potentially critical practice as it creates a public awareness for the hierarchies of urban spaces after dark, where illuminations often reflect the privileges of those who already live on the sunny side of the street and reproduce the marginalisation of social groups who are used to living in the shadow. It is thereby important to note that 'dark' no longer equals 'poor' as it used to in the past (Bouman 1987; Entwistle et al. 2015). Instead, in a world of abundant lighting, the dividing line runs between well-designed and poorly designed illuminations.

Figure 2.4
Two young participants
of the 2011 Guerrilla
Lighting event in
Sclessin, Liège.
*Photo: Nona
Schulte-Römer.*

Figure 2.5
Explorations into the
'epistemic wasteland':
NightSeeing events
with New York
lighting designer Leni
Schwendiger (in the
centre) are another
eye-opening format.
*Photo: Nona
Schulte-Römer.*

Conclusion

In his book *Experience*, Scott Lash differentiates between 'objective
experience' as the basis of modern (scientific) knowledge production and subject-object thinking, and 'subjective experience' as lived

experience (Lash 2018: 1–2). With regard to the experience of light in public spaces, this differentiation is reflected in different epistemic approaches. While the 'objective' experience is key to knowledge production with a focus on reproducible, generalizable cause-effect mechanisms between observer and light, the subjective experience can produce knowledge about light that is situated and embodied. With industrialisation, the objectified approach to lighting became dominant and 'objective' scientific measures the evidence base for functional street and area lighting, while decorative illuminations and aesthetic experiences became the realm of subjective experiences. In this chapter, I have described how this differentiation of epistemic approaches coincided with a fragmentation of lighting purposes, practices and perceptions. While 'objective' knowledge derived from photometric experiments and statistical approaches shapes lighting standards to the present day, subjective experiences are more difficult to grasp as they involve a co-production of sensuous light and sensing individuals, of affect and of atmospheres (Edensor 2015; Böhme 1995; Bille 2019; Sumartojo et al. 2019).

In recent years, subjective experiences of light in public spaces have received increasing scholarly attention by anthropologists, sociologists and geographers. This edited volume is yet more proof of this ongoing exploration of what once appeared as an 'epistemic wasteland' (Hasse 2007). In this chapter, I have argued that not only social scientists, but also lighting designers can contribute to this exploration.

Understood as an epistemic culture (Knorr-Cetina 1999), professional lighting design takes specific lighting situations as its evidence base, often without discriminating between functional and decorative lighting. This professional vision (Goodwin 1994) is coupled with scientific insights into the physical, physiological and psychological effects of artificial light, standards and regulations, electrical engineering, and increasingly also, electronics. This augmented 'objective' perspective on light and lighting designs thus acknowledges the adverse effects of light on wildlife and humans, but also reframes them in terms of a subjective experience – the loss of darkness and the starry night sky (Narboni 2009; Edensor 2017; Schulte-Römer et al. 2018: 47). Finally, the cultivation of engagement formats for participatory design, such as public night walks, workshops or experimental Guerrilla Lighting interventions, create a space in lighting design for enacting the social (Slater et al. 2018). They allow lighting designers and laypersons to enter in an exchange with each other and find a shared language to express and discuss both subjective experiences and objectified professional visions.

To conclude, the exploration of lived subjective experiences of light in public spaces seems a good basis for reinventing 'the social' in lighting – beyond generically 'safe' public spaces or spectacular aestheticised events and 'place-branding iconicity' (Entwistle and Slater 2019). Today, we find ourselves in a situation where the 'LED revolution' challenges taken-for-granted lighting practices (cf. Seitinger 2010), while an increasing awareness for climate change and light pollution calls for a renegotiation and reconsideration of lighting needs – beyond unquestioned cultural desires for more light. In this situation, it seems that lighting design has developed professional practices that are suited to redrawing boundaries and support sustainable lighting – by acknowledging not only efficiency criteria but also social and ecological aspects, and by promoting a situated way of seeing, an interdisciplinary and qualitative mode of evidence production, and transdisciplinary engagement of affected parties.

Of course, I am depicting an ideal world. In reality, lighting designers are not exempt from the economic pressures that cause bad-quality lighting all over the world. If the budget is smaller than expected, lighting design is often the first item on the bill that gets cancelled (Schulte-Römer et al. 2018: 141). Moreover, boundary work in lighting design is ongoing in these times of sociotechnical transition towards more sustainable LED lighting. Although the previously described approaches and initiatives receive positive feedback during conferences and in the lighting field, they do not automatically become mainstream and transform the field.

The exploration of the complex subjective experiences of light in public spaces seems nevertheless an important prerequisite when it comes to planning light in more reflexive, democratic and, hence, sustainable way. After all, sustainable development in lighting entails more than energy and cost savings, light pollution mitigation, and 'light for the people'. If we follow the seminal definition of the Brundtland report (1987), the greater challenge seems the development of sustainable planning processes that allow the multiple stakeholders of urban lighting to negotiate present and future needs for light and darkness in the contexts of technological development, resource exploitation, unequally distributed resources and institution-alised power asymmetries in the lighting field.

Notes

1. The workshop was organised in the context of our research program 'Cultural Source of Newness' (Hutter et al. 2010) and hosted together with my colleague Ariane Berthoin Antal at WZB – Berlin Social Science Center. The expert workshop participants were three lighting designers, Ulrike Brandi (Ulrike Brandi Licht – Hamburg, Germany), Isabelle Corten (Radiance 35 – Liège, Belgium) and Roger Narboni (Agence Concepto – Paris, France); geographers Jean-Michel Deleuil (Urban Development Department, INSA – National Institute of Applied Sciences – Lyon) and Steve Millington (Manchester Institute of Social & Spatial Transformations, Manchester Metropolitan University); municipal urban planning and lighting experts Alexandre Colombani (LUCI – Lighting Urban Community International – Lyon), Cathy Johnston (Development and Regeneration Services, Glasgow City Council) and Lars Löbner (Public Space Design, City Planning Office, City of Leipzig); as well as three representatives of the lighting industry and Joachim Leibig (Hess – Villingen-Schwenningen), Peter Uhrig (Se'lux – Berlin). The workshop report (Schulte-Römer 2010) has not been published but is available on demand.
2. Entwistle and Slater (2019) refer to technical and aesthetic lighting. I chose the term 'functional' as this is a commonly used expression among lighting professionals and 'decorative' in order to avoid confusion with what I describe as 'aesthetic' – meaning, a reflexive engagement with sensuous experiences of light (Hirdina 1997).
3. The lighting designers were Elettra Bordonaro (Italy), Isabelle Corten (Belgium), Jöran Linder and Eric Olsson (Sweden), and Martin Lupton and Sharon Stammers (Great Britain).

References

Barnaby, Alice. 2009. "Light Touches: Cultural Practices of Illumination, London 1780–1840." Doctoral thesis, Exeter: University of Exeter. http://hdl.handle.net/10036/3037.

Besecke, Anja, and Robert Hänsch. 2015. "Residents' Perceptions of Light and Darkness." In *Urban Lighting, Light Pollution and Society*, edited by Josiane Meier, Ute Hasenöhrl, Katharina Krause and Merle Pottharst. New York: Routledge.

Bille, Mikkel. 2019. *Homely Atmospheres and Lighting Technologies in Denmark: Living with Light*. London: Bloomsbury Publishing.

Bille, Mikkel, and Tim Flohr Sørensen. 2007. "An Anthropology of Luminosity: The Agency of Light." *Journal of Material Culture* 12(3): 263–284.

Binder, Beate. 1999. *Elektrifizierung als Vision: zur Symbolgeschichte einer Technik im Alltag*. Tübingen: Tübinger Vereinigung für Volkskunde.

Böhme, Gernot. 1995. *Atmosphäre. Essays zur neuen Ästhetik*. Frankfurt am Main: Suhrkamp.

Bouman, Mark J. 1987. "Luxury and Control." *Journal of Urban History* 14(1): 7–37. doi: 10.1177/009614428701400102.

Brox, Jane. 2010. *Brilliant: The Evolution of Artificial Light*. Boston; New York: Houghton Mifflin Harcourt.

Brundtland, Gro Harlem, Mansour Khalid, Susanna Agnelli, Sali Al-Athel, Bernard Chidzero, Lamina Fadika, Volker Hauff, Istvan Lang, Ma Shijun, and Margarita Morino de Botero. 1987. Our Common Future. In *Brundtland Report*. New York: United Nations.

Chain, Cyril. 2010. *Norme NF EN 13201 – Enquête sur son utilisation et sa révision*. Paris: Certu – Ministère de l'Écologie, du Développement et de l'Aménagement durables.

Challéat, Samuel, Dany Lapostolle, and Rémi Bénos. 2015. "Consider the Darkness. From an Environmental and Sociotechnical Controversy to Innovation in Urban Lighting." *Journal of Urban Research* 11: 1–17. doi: 10.4000/articulo.3064.

Dannemann, Etta. 2017–11–09. "Meaningful Light – How Colored Landmark Lighting is Becoming a Political Issue." Berlin: Stadt nach Acht.

Ebbensgaard, Casper Laing. 2019. "Making Sense of Diodes and Sodium: Vision, Visuality and the Everyday Experience of Infrastructural Change." *Geoforum* 103: 95–104. doi: 10.1016/j.geoforum.2019.04.009.

Ebbensgaard, Casper Laing. 2020. "Standardised Difference: Challenging Uniform Lighting through Standards and Regulation." *Urban Studies* 57(9): 1957–1976. doi: 10.1177/0042098019866568.

Edensor, Tim. 2012. "Illuminated Atmospheres: Anticipating and Reproducing the Flow of Affective Experience in Blackpool." *Environment and Planning D: Society and Space* 30(6): 1103–1122. doi: 10.1068/d12211.

Edensor, Tim. 2015. "Light Design and Atmosphere." *Visual Communication* 14 (3): 331–350.

Edensor, Tim. 2017. *From Light to Dark: Daylight, Illumination, and Gloom*. Minneapolis, MN: University of Minnesota Press.

Edensor, Tim, and Mikkel Bille. 2017. "'Always Like Never Before': Learning from the Lumitopia of Tivoli Gardens." *Social & Cultural Geography* 1–22. doi: 10.1080/14649365.2017.1404120.

Edensor, Tim, and Steve Millington. 2013. "Blackpool Illuminations: Revaluing Local Cultural Production, Situated Creativity and Working-Class Values." *International Journal of Cultural Policy* 19(2): 145–161.

Entwistle, Joanne, and Don Slater. 2019. "Making Space for 'The Social': Connecting Sociology and Professional Practices in Urban Lighting Design 1." *The British Journal of Sociology*. doi: 10.1111/1468-4446.12657.

Entwistle, Joanne, Don Slater, and Mona Sloane. 2015. "Darkness has become a Luxury Good in London": On the Social Meaning of Street Lighting." *CityMetric*. 2017. 20 October 2015. http://www.citymetric.com/horizons/darkness-has-become-luxury-good-london-social-meaning-street-lighting-1504, accessed 13 December.

Gieryn, Thomas F. 1983. "Boundary-Work and the Demarcation of Science from Non-Science: Strains and Interests in Professional Ideologies of Scientists." *American Sociological Review* 48(6): 781–795. doi: 10.2307/2095325.

Gonzalez, Edna Hernandez. 2010. *Comment l'illumination nocturne est devenue une politique urbaine: la circulation de modèles d'aménagement de Lyon (France) à Puebla, Morelia et San Luis Potosí (Mexique)*. Doctoral thesis. Paris: Université Paris-Est. https://tel.archives-ouvertes.fr/tel-00601294/.

Goodwin, Charles. 1994. "Professional Vision." *American Anthropologist* 96(3): 606–633.

Green, Judith, Chloe Perkins, Rebecca Steinbach, and Phil Edwards. 2015. "Reduced Street Lighting at Night and Health: A Rapid Appraisal of Public Views in England and Wales." *Health & Place* 34: 171–180. doi: 10.1016/j.healthplace.2015.05.011.

Hansen, Ellen Kathrine, and Michael Mullins. 2014. "Lighting Design-Toward a Synthesis of Science, Media Technology and Architecture." *Smart and Responsive Design* 2(32): 613–620.

Hasenöhrl, Ute. 2015. "Lighting Conflicts from a Historical Perspective." In *Urban Lighting, Light Pollution and Society*, edited by Josiane Meier, Ute Hasenöhrl, Katharina Krause and Merle Pottharst, 105–124. New York: Routledge.

Hasse, Jürgen. 2007. "Das künstliche Licht in der Architektur – eine epistemische Brache." In *Die Alte Stadt*, edited by Jürgen Hasse, 67–77. Remshalden: BAG-Verlag.

Hirdina, Karin. 1997. *Belichte. Beleuchten. Erhellen*. Edited by Humboldt-Universität zu Berlin. Vol. 89, *Öffentliche Vorlesung*. Berlin: Humboldt-Universität zu Berlin, Humboldt-Universität zu Berlin, Philosophische Fakultät III, Institut für Kultur- und Kunstwissenschaften, Seminar für Ästhetik.

Hölker, Franz, Timothy Moss, Barbara Griefahn, Werner Kloas, Christian C Voigt, Dietrich Henckel, Andreas Hänel, Peter M. Kappeler, Stephan Völker, and Axel Schwope. 2010. "The Dark Side of Light: A Transdisciplinary Research Agenda for Light Pollution Policy." *Ecology and Society* 15(4).

Hoormann, Anne. 2003. *Lichtspiele: zur Medienreflexion der Avantgarde in der Weimarer Republik*. München: Wilhelm Fink Verlag.

Hutter, Michael, Ariane Berthoin Antal, Ignacio Farías, Lutz Marz, Janet Merkel, Sophie Mützel, Maria Oppen, Nona Schulte-Römer, and Holger Straßheim. 2010. Research Program of the Research Unit "Cultural Sources of Newness". In *Discussion Paper*. Berlin: WZB.

Isenstadt, Sandy. 2018. *Electric Light: An Architectural History*. Cambridge, MA: MIT Press.

Jakle, John A. 2001. *City Lights: Illuminating the American Night*. Baltimore and London: Johns Hopkins University Press.

Johnston, Sean François. 1994. "A Notion or a Measure: The Quantification of Light to 1939." Unpublished PhD dissertation, Leeds: University of Leeds.

Knorr-Cetina, Karin. 1999. *Epistemic Cultures: How the Sciences Make Knowledge*. Cambridge, MA: Harvard University Press.

Koslofsky, Craig. 2011. *Evening's Empire: A History of the Night in Early Modern Europe*. Cambridge: Cambridge University Press.

Lamont, Michèle, and Virág Molnar. 2002. "The Study of Boundaries in the Social Sciences." *Annual Review of Sociology* 28: 167–195.

Lash, Scott. 2018. *Experience: New Foundations for the Human Sciences*. Hoboken, NJ: John Wiley & Sons.

Latour, Bruno, and Emilie Hermant. 2006 [1998]. Paris: Invisible city [Paris ville invisible]. edited by Translated from the French by Liz Carey-Libbrecht and Corrected by Valérie Pihet.

Light Collective. 2008. "Guerrilla Lighting." https://lightcollective.net/light/ing/guerrilla-lighting, accessed July 2021.

McQuire, Scott. 2008. *The Media City: Media, Architecture and Urban Space*. London: Sage.

Meier, Josiane. 2019. "By Night. An Investigation into Practices, Policies and Perspectives on Artificial Outdoor Lighting." Doctoral thesis. Berlin: Technische Universität Berlin. https://depositonce.tu-berlin.de/handle/11303/9199.

Melbin, Murray. 1987. *Night as Frontier: Colonizing the World after Dark*. New York: Free Press.

Morgan-Taylor, Martin. 2015. "Regulating Light Pollution in Europe: Legal Challenges and Ways Forward." In *Urban Lighting, Light Pollution and Society*, edited by Josiane Meier, Ute Hasenöhrl, Katharina Krause and Merle Pottharst, 159–176. New York: Routledge.

Narboni, Roger. 2009. *La nuit disparue*. Florence: Targetti Foundation Publishers.

Narboni, Roger. 2021. Trame noire – Le temps de la maturité. *LUMIÈRES* 35(Juin): 44–47.

Nye, David E. 2018. *American Illuminations: Urban Lighting, 1800–1920*. Cambridge, MA: MIT Press.

Otter, Chris. 2008. *The Victorian Eye: A Political History of Light and Vision in Britain, 1800–1910*. Chicago: University of Chicago Press.

Pérez Vega, Catherine, Karolina M. Zielinska-Dabkowska, and Franz Hölker. 2021. "Urban Lighting Research Transdisciplinary Framework – A Collaborative Process with Lighting Professionals." *International Journal of Environmental Research and Public Health* 18(2): 624.

Pinch, Trevor. 2010. "On Making Infrastructure Visible: Putting the Non-Humans to Rights." *Cambridge Journal of Economics* 34(1): 77–89.

Reckwitz, Andreas. 2009. "Die Selbstkulturalisierung der Stadt. Zur Transformation moderner Urbanität in der 'creative city'." *Mittelweg 36*(2): 2–34.

Schivelbusch, Wolfgang. 1988. *Disenchanted Night. The Industrialisation of Light in the Nineteenth Century*. Translated by Angela Davies. Oxford, UK: Berg.

Schulte-Römer, Nona. 2010. *Urban Light Planning – Evaluation, Evidence and the New – Unpublished Workshop Report*. Berlin: WZB – Social Science Research Center.

Schulte-Römer, Nona. 2011. "Enlightened Cities. Illuminations for Urban Regeneration." In *Understanding the Post-Industrial City*, edited by Frank Eckardt and Sofia Morgado, 128–165. Würzburg: Königshausen & Neumann.

Schulte-Römer, Nona. 2015. "Innovating in Public. The Introduction of LED Lighting in Berlin and Lyon." Doctoral thesis. Berlin: Technical University Berlin. https://depositonce.tu-berlin.de/handle/11303/5211.

Schulte-Römer, Nona, Etta Dannemann, and Josiane Meier. 2018. *Light Pollution – A Global Discussion*. Leipzig: Helmholtz-Centre for Environmental Research GmbH – UFZ.

Schulte-Römer, Nona, Josiane Meier, Etta Dannemann, and Max Söding. 2019. "Lighting Professionals versus Light Pollution Experts? Investigating Views on an Emerging Environmental Concern." *Sustainability* 11(6): 1696. doi: 10.3390/su11061696.

Seitinger, Susanne. 2010. *Liberated Pixels: Alternative Narratives for Lighting Future Cities*. Doctoral thesis. Boston, MA: Massachusetts Institute of Technology. http://hdl.handle.net/1721.1/61935.

Shaw, Robert. 2015. "Night as Fragmenting Frontier: Understanding the Night that Remains in an Era of 24/7." *Geography Compass* 9(12): 637–647.

Slater, Don, Elettra Bordonaro, Joanne Entwistle, and Isabelle Corten. 2018. "Social Lightscapes Workshops: Social Research in Design for Lighting Professionals." London: London School of Economics and Political Science. http://eprints.lse.ac.uk/87187/.

Social Light Movement. 2011. "1st Social Light Movement Workshop in Sclessin, Belgium." https://sociallightmovement.com/missions/workshop/sclessin/, accessed July 2021.

Star, Susan Leigh, and Karen Ruhleder. 1996. "Steps Toward an Ecology of Infrastructure: Design and Access for Large Information Spaces." *Information Systems Research* 7(1): 111–134.

Sumartojo, Shanti, Tim Edensor, and Sarah Pink. 2019. "Atmospheres in Urban Light." *Ambiances. Environnement sensible, architecture et espace urbain* (5).

Tomory, Leslie. 2009. "Progressive Enlightenment: The Origins of the Gaslight Industry 1780–1820." Ph.D., Institute for the History and Philosophy of Science and Technology, University of Toronto.

Wings professional studies. 2021. *Lighting Design Master Program*. Wismar University of Applied Sciences. www.wings-university.com/lighting_design, accessed August 2021.

Zerubavel, Eviatar. 2015. *Hidden in Plain Sight: The Social Structure of Irrelevance*. New York, NY: Oxford University Press.

Zielinska-Dabkowska, Karolina, and Michael F. Rhode, eds. 2017. *New Perspectives on the Future of Healthy Light and Lighting in Daily Life*. Wismar: callidus. Verlag wissenschaftlicher Publikationen.

Chapter 3

LIGHT AND VALUE

A design anthropology of light and well-being in hospital buildings

Sarah Pink, Melisa Duque, Shanti Sumartojo and Laurene Vaughan

DOI: 10.4324/9781003182610-3

Introduction

The meaning and use of light are always contingent, and it is impossible to predict precisely how light will be experienced in any given situation. This makes lighting design tricky, and in this chapter, we explore what an anthropological conceptualisation of value can contribute to this question. We discuss the relationship between light and value by examining two questions. First, we draw on our ethnographic research to ask how people's engagements with light are expressed as having value in everyday situations on the one hand and, on the other, how feelings about light can express cultural or collective values. Second, we reflect on the significance of these findings for lighting design. To develop our discussion, we draw on examples based on our ethnographic research, which demonstrate how light was distributed, used and discussed in the context of a newly built hospital. The role of daylight in generating staff well-being in workplace environments has been well documented in academic and applied disciplines that focus on the design of the built environment. However, while our research supported this existing research in the field of environmental health design that suggests that natural light has well-being benefits, it showed how the experience of natural versus artificial lighting in hospital environments is also contingent on professional values. We therefore advocate for a design anthropological approach to light, which invites us to consider how the experience and meaning of light is driven not only by good design but by the values and modes of innovating to create value that are integral to workplace cultures and practices.

The hospital environment provides an ideal example through which to examine these questions. Our study was undertaken in the psychiatric units of a large hospital in regional Victoria, Australia. The new hospital building, which has won several awards, has an impressive commitment to bringing natural light into the lives of staff and patients, particularly in its atrium and, for instance, through a successful internal courtyard system within the building.

The introduction of significant natural light and new modes of electric lighting in the new hospital's psychiatric units was also very successful and provides insights both into how lighting matters as well as how and where future design might learn from examples of where light and other social and material elements configured best. This context, moreover, reveals how resources of light and lighting

Figure 3.1
Views to the outside landscape from the psychiatry reception.
Photo: Melisa Duque.

Figure 3.2
The light-filled courtyard.
Photo: Melisa Duque.

participate in how hospital staff experience built environments in ways that are inflected by the specificity of their everyday work-place circumstances and professional values. This makes it possible to explore the contingencies that shape the relationship between the presence of daylight in the built hospital environment and staff well-being, and to explore the locally and culturally specific variables that are often otherwise hidden under the surface. In doing so, we develop and demonstrate an approach to understanding how daylight contributes to the ways people 'feel' in buildings through attention to social, cultural and experiential dimensions informed by design anthropological theory.

Investigating the experience of light through concepts of value offers a perspective on the question of how and why light participates in everyday meaning making. As we explain later, anthropologists of light understand its meaning and uses as being inextricable from individual, social, cultural and environmental things and processes that life entails (Pink and Leder Mackley 2016; Pink and Sumartojo 2017; Bille 2019). That is, the way we sense and engage with light has to do with how we live and is part of everyday worlds. We are interested in how light is associated with value and values as a mode of investigating how it is at once bound up with the measurement and audit-based structures of neoliberal society, and the sensory, non-representational and often invisible feelings through which the everyday is lived. As the anthropologist David Graebner explains, the relationship between value and values has figured strongly in anthropological value theory since the 1980s. He points out that it is not a coincidence that 'we use the same word to describe the benefits and virtues of a commodity for sale on the market (the "value" of a haircut or a curtain rod) and our ideas about what is ultimately important in life ("values" such as truth, beauty, justice)'. Yet he argues, 'There is some hidden level where both come down to the same thing'. Graebner explains this with reference to the difference between commoditised labour, where a good or service is valued in relation to the cost of the labour that produces it, and unpaid labour, such as housework. The former is given a commodity value, and the latter is driven by social or cultural values; but both are, in fact, forms of labour (Graeber 2013: 224).

In this chapter, a similar logic is used to distinguish among a series of different ways in which light comes together with value and values. These can be summed up as follows:

- When light is given value through quantitative measurement techniques, which do not account adequately for human experience.

- When the value of light is accounted for in relation to the assumption that light can impact *on* people.
- When people express their experiences of light qualitatively, in ways that account for how these experiences are valued.
- When people explain their experiences of light through the filters of specific cultural or professional values.
- When in situations where these different ways of measuring and evaluating light converge in the same professional environment.

We return to these modes in our discussion later of how light is experienced in hospital environments to demonstrate how they play out. First, in the next sections, we examine how light, lighting design and healthcare environments have been brought together in recent literature in the social sciences, healthcare environments and design research.

Light and the experience of healthcare environments

There have recently been calls for further attention to the complexities and the qualitative dimensions of human experience of the healthcare built environment, including recent interest in the sensory modes in which people experience built hospital environments in the medical humanities (Bates 2018), qualitative social sciences (Buse et al. 2018; Martin et al. 2015; Harris, Rice) and architecture studies (Annemans et al. 2017). Much of such literature focuses on patient experience, as it does in more quantitative approaches to health environments, and Martin et al. note the 'large body of evidence that points to the role of the designed environment in the efficacy of care: floor layouts, noise levels, lighting, single rooms, ventilation, exposure to daylight, access to 'green' environments, proximity to windows and so on have all been found to impact upon health outcomes' (Martin et al. 2015: 1008). Dijkstra et al. (2006), in an earlier review of literature about the impact of features such as sunlight, on patient outcomes, similarly found that sunlight appears to speed recovery in several cases but warn that some conditions are worsened by sunlight, thus recommending care and contextual evaluation. Other relevant literature on the environmental quality of hospitals, which accounts for hospital staff, concurs regarding the benefits of light. For instance, Dalke et al. argue that 'The visual environment, including quality of daylight and electric light, is a vital element influencing hospital staff morale and productivity' (2006: 344) and there has been a continued emphasis on the benefits of natural light for both staff and patient well-being, advocating, in some cases, user-centred design (Huisman et al. 2012; Shikder et al. 2012; Andrade et al. 2012). In this field of research, such arguments are primarily based on quantitative questionnaire studies or surveys (see Iyendo et al. 2016) endorsed by 'environmental quality perception (EQP)' measurement techniques that attend to 'cognitive-psychological processes involved in the evaluation of environmental qualities' (Andrade et al. 2012: 99) and 'Perceived Hospital Environment Quality Indicators (PHEQIs)' (Andrade et al. 2012:100) or combine interviews and quantitative measures (e.g., Nejati et al. 2016). These methods create systematic studies of relatively larger numbers of participants than would be accounted for by qualitative research and are important in demonstrating and endorsing the role of light in the generation of staff and patient well-being. Yet they do not offer in-depth understanding of the social, cultural and experiential dimensions of how and why hospital staff and patients experience and work with light in particular ways.

While there is a nascent interest in anthropology of lighting design in industry contexts[1] (which we refer to later), in studies of lighting technologies, likewise, there tends to be a quantitative focus on how people react to light. The most common approach is to place those technologies at the centre of inquiry, focusing on what their affordances are and how they interact with their surroundings. This includes work that measures brightness, diffusion and the bounce of light onto surfaces, or that evaluates colour and its effect on visual perception (e.g., Boyce 2014). To evaluate how such effects feel to people that encounter them, scales or surveys are used to rate lighting conditions along scales of what feels 'safe' or 'pleasant' (Knight 2010) but, crucially, do not allow participants to qualify what these terms mean for them or nominate their own categories of feeling or sensory experience. Such studies, precisely because of their orientation to the technologies of lighting design, account for their effects in terms that are defined by and structured through the technologies. In other words, ways of experiencing light that do not directly address their technological aspects may not be accounted for or understood by researchers, or even communicated by participants. Aspects of lighting technology thus provide the sole terms for understanding light in this approach, which does not reach beyond this to probe the subtler, emotional or experiential qualities of light. As the anthropologist of light Mikkel Bille reports, 'A large amount of research has studied how the physiological body and individual psychology reacts to various light, and its physical, spatial and environmental properties, mostly through positivist experiments and measurements' and 'in human factors approaches to light design there is a lack of attention to the social and cultural aspects of light'. As Bille puts it, in such approaches, it is 'as if cultural traditions and social life are unknowable, or secondary to the individual bodily response to light, rather than embedded in the very way people use, sense and make sense of light' (Bille 2019: 6). Indeed, it has been argued that lighting design in particular – which by its nature can seep, infuse, dissipate or mingle with other aspects of our surroundings – needs to attend to the messy and ongoing nature of experience, its moods and atmospheres, and its material, sensory and affective qualities (Jensen et al. 2018).

Thus, if we take together this gap in knowledge relating to the social, cultural and experiential qualities of light in designing healthcare environments and in lighting design in general, a common gap in knowledge and practice appears. While this gap has been revealed in the research of social scientists, the writings of designers themselves also throw doubt on the capacity of design to alone resolve problems in public service organisations, such as mental health care or in user experiences of lighting, without the help of anthropologists. Two recent commentaries, in particular, urge for a greater presence of anthropologists in both fields of design practice. In a 2018 *Design Issues* article the designer Pierri criticises 'articles, publications, and case studies that have been selling the role of design in public and community organizations as the panacea to all problems'. She complains that they 'reproduce a discourse about design for social change that tells a single story of design, in which the latter is usually uncritically depicted as an effective way to understand the human experience, to increase creativity, to quickly learn "by doing" and "by failing fast", to engage stakeholders in collaborative efforts, and to revolutionize public services and contribute to opening up bureaucracies' (Pierri 2018: 25). She thus asks, 'The question that comes to mind, when reading these accounts of design, is this: When and where did design build the credibility and provide the credentials to do this demanding job?' (Pierri 2018: 26) and proposes, in the context of doing design research with mental health services users – as co-researchers – alternative modes of participating as a designer in mental health contexts, which draw from anthropological practice. Also in 2018, Jose Manuel Dos Santos, an industrial designer and head of design and user experience (Americas) at Signify, the world's leading company in conventional and LED lighting, commented that 'I still think designers need to work with anthropologists that understand the needs of designers in a project/ professional context'. In a similar vein to Pierri, Dos Santos noted that 'Design itself has been accused of wanting to do more and offer more than it is capable of, diluting its core offer and doing a sloppy job at things that end up weakening their credibility' (https://blog.antropologia2-0.com/en/design-anthropology-at-phillips/).

Design anthropology offers a particular way forwards that, we argue, can fill these gaps in knowledge and practice (see also Smith et al. 2016; Pink et al. 2017). It draws from a robust anthropological theory of light and lighting, and its relationship to the built environment, interiors and human relationships, in-depth ethnographic understandings of everyday circumstances in which, and feelings through which, light is experienced and used. We outline the significance of the anthropology of light and lighting further later.

Anthropologies of light and lighting

In earlier writing, we have advanced the concept of the 'lit world' (Pink and Sumartojo 2017). The 'lit world' concept, inspired by the anthropologist Ingold's (2010: 122) notion of the 'weather world', suggests that we do not simply perceive light as something that is separate from us but that we 'perceive in' light (Ingold 2010: 131). The work of Bille adds to this in treating light not only as part of the environment but as a technology, whereby ' "Light for", . . . highlights how technology is intrinsically embedded in human perceptions, practices and meaning-making'. This he proposes, 'leads to a more human-relevant approach in which the relationship between the sentient embodied subject and the world emerges'. He concludes that 'People thus live with light and not just in it' (Bille 2019: 9). In more practical terms, these ways of living with and in light are exemplified in recent studies. Based on research that showed how people navigated their night-time homes, using it on their way to bed and using indirect sources of light – from other rooms, from outdoors and from their smartphones – Pink and Leder Mackley (2016: 179) have argued that 'Here light is experienced, rather than contemplated' in that people see with light, rather than seeing or visually contemplating light itself. This is significant for what we consider light to be and what our relationship to it is. Pink and Leder Mackley emphasise the arguments of Ingold and the geographer Tim Edensor that light is not 'a phenomenon of the physical world . . . or of the interior mind' but is 'an experience' (Ingold 2011: 128), and that 'We see both with and in light and move through illuminated space, making engagement with light a deeply embodied experience' (Edensor 2012: 1107). As Pink and Leder Mackley emphasise, here, light is experienced in movement through environments of which light is part, an ongoingly emergent and shifting mode of experiencing. As such, light itself is not fixed or static but ongoingly emergent within a similarly emergent world. It is experienced by people in movement, even if sitting or lying still, moving on through the temporality of the world. Indeed, as Bille has argued, with reference to lighting in homes, 'beyond offering visibility, lighting is a practice of attuning atmospheres for a variety of activities to take place in the home, such as social gatherings, cooking, relaxing, entertaining, cleaning, and feeling secure' (Bille 2019: 8). Hospital units are, equally, places of activities and feelings; and as the examples we discuss later highlight, there are different modes of practice, activity and feeling that might be experienced and played out in relation to light.

In this chapter, we are interested, with reference to the particular example of staff working in psychiatric units, in how people inhabit and work in and with light, and in doing so, how they not only see with light but feel with light, and how this is associated with value and values. In our research and analysis, we are concerned with lighting design in the sense of the design and experiences of buildings and windows and of artificial lighting systems. In doing so then, we appreciate the distinction 'between "light" (a material form emanating from a natural or artificial source reflecting, changing or absorbing on material surfaces and an object of scientific scrutiny), and "lighting" (as a practice of manipulating, designing and appreciating luminosity in itself or to inform other practices)' (Bille 2019: 18). However, we emphasise that natural as well as artificial light are engaged in architectural and interior design and the relational ways that they are experienced.

The approach we have advanced earlier for understanding light and lighting as part of an ongoingly emergent world is coherent with the design anthropological approach that forms the basis of our methodology. Design anthropology has developed in various iterations, and the focus with which our work is most closely aligned leans towards phenomenological anthropology and participatory design (Gunn et al., Smith et al. 2016). This branch of design anthropology is underpinned by the concept of emergence, which refers to the ongoing and continually changing (albeit sometimes in unnoticeable ways) environments in which we live and participate (Smith and Otto 2016). Through a focus on emergence design, anthropologists are able to acknowledge the states of uncertainty and possibility (Akama et al. 2018) in which people live and the unfinishedness of the things and processes that we encounter in research and that we design. Here, building on insights from engaging design anthropology in the research of built environments (Pink et al. 2017) and technology (Pink et al. 2018), we tailor an approach to light in hospitals. In doing so, we are particularly concerned with the concept of contingency, which reminds us that the emergent circumstances of everyday life are composed of specific configurations of things and processes that might not immediately be visible until researched (e.g., Irving 2017). By acknowledging the contingent circumstances in which people always find themselves but only tend to notice when the circumstances are unusual, design anthropologists can gain understandings of what different elements need to come together to create circumstances where people will feel familiar and confident. During our research in the hospital psychiatric units, this concept drew our attention to how staff experiences of lighting or feelings about the need for particular lighting in specific areas or moments was contingent on other factors, which could be material, emotional or concerned with values or ethics. Later, we bring these design anthropological concepts together with those of value and values in dialogue with ethnographic research designed specifically to provide under-the-surface understandings of human activity and the feelings concerned.

Context and methods

To develop our approach and argument, we draw on a three-year research project undertaken between 2016 and 2018. Our research investigated the sensorial and lived experience of patients, staff and visitors to the psychiatric units at the Bendigo Hospital in Victoria, Australia. We undertook ethnographic research in three separate existing psychiatric units in 2016. In 2017–2018, these units moved to a new purpose-built facility where they were co-located within the larger hospital. These units were the Adult Acute Unit (AAU), Older People Unit (OPU) and Extended Care Unit (ECU) for long-term patients. In the new hospital, they were joined by a new Parent and Infant Unit (PIU), and the AAU and OPU were divided into Low- and High-Dependency Units (LDU and HDU, respectively). Moreover, 150 people participated in our research. All participants gave informed consent.

Our wider study focused on the transition to the new facility and the impact of the new design on staff and patient well-being. It included attention to sensory features of the design, including the experience of light and lighting. This provides an ideal example through which the experience of light in healthcare settings is examined, since although our examples focus exclusively on psychiatric units, the study as a whole has covered seven different old and new environments, and includes a range of different patient groups and the staff who care for them, ranging between secure high-dependency units, where there is an enduring possibility of violence, to low-dependency older people, dementia patients, and a parent and infant unit, considered by staff to be the least stressful of the environments. Thereby, this enabled us to examine how light and lighting similarly configured as part of the new hospital environment and, thus, became part of the experience of different staff and patient groups. In this article, we focus primarily on the experience of staff, which in part we do to redress what is seen as an imbalance in attention to well-being experiences of hospital staff in favour of patient experience noted earlier. However, in the case of the psychiatric units, which are considered to be stressful workplaces, we have also found that the ways that staff experience and use light offers particularly interesting examples of how light emerges as part of a healthcare workplace experience and culture, as specific to the cultures and socialities of psychiatric care work in Australia. These are supported by a series of further insights into the wider context of feelings associated with light and lighting in the hospital.

Our approach draws on sensory ethnography (Pink 2015), which was originally developed largely for researching in the built environment (Pink 2004, 2012; Pink et al. 2017), making it particularly suitable for investigating the experience of hospital environments. Sarah Pink developed the sensory ethnography research design, which was underpinned by teamwork. Shanti Sumartojo undertook the first stage of fieldwork during ten weeks in 2016. Stage 2 was undertaken by Melisa Duque over ten months between 2017 and 2018, accompanied by Pink when possible. We developed sensory ethnography interviewing, touring and re-enactment techniques where participants showed us and commented on environments and feelings, or demonstrated activities, did ethnographic 'hanging around' as well as reflected on our own experiences of being in and sensing the 'feel' of the fieldwork sites (Pink 2015). The immersed researchers worked with Pink during the fieldwork and analysis to collectively build dialogue among the experience of immersed researchers, among theoretical and interpretive insights, and additionally, with Laurene Vaughan, to connect this with design and concepts of care.

The lit hospital

In this section, we discuss the hospital as a kind of 'lit world', explained at the beginning of this article, whereby in this case, people inhabit and experience its built environment in and with both natural and artificial light, both of which form part of the hospital building and interior design. Such light participates in configuring certain circumstances of well-being, it is implicated in how we see and feel healthcare environments, and the experience of it is inflected with human values.

The new Bendigo Hospital, including the psychiatric units, was designed to optimise the presence of natural light. The reception area where people pass through when entering the department is on the second floor and opens to the large light-filled atrium at the heart of the hospital building. Within the psychiatric department, each unit has access to one or more open courtyard spaces, which is separated from the indoor areas by large glass windows, allowing light to flow into the communal areas and corridors internal to the units. Additionally, the common areas and patient bedrooms have large external windows that are expansive sources of daylight. Staff, patients and visitors were consistently enthusiastic about the natural light of the units, as well as the artificial lighting in patient rooms and other areas in the units. In what follows, we focus on the complexities of how this light was implicated in the ways staff felt.

Light as valued

Staff were consistently impressed by and enthusiastic about the light-filled entrance to the hospital, and as they moved through, participants felt and valued aspects of the environment that they were part of. Here are the words of one participant:

> I think the thing that I like very much about the new hospital is the stairwell coming up with the big open windows. So you immediately feel like you're inside and outside at the same time. You can be engaged with the outside world, whereas in a lot of hospitals you can feel like you're in an office and you're a little bit isolated from what's happening outside. You can see the weather outside. You can see the sun and all the people coming and going. I think that's really good. So I try and use the stairs as much as possible.

> (Consultant)

The daylight that entered the new units was discussed comparatively in relation to the old hospital. Here, the way the new sense of safety, welcome and openness that the staff experienced in these daylit spaces was valued was expressed in the staff's discussions of their experiences in both old and new hospitals. For example, here is how one staff member described it:

> I really love the unit in regards to it's a much brighter welcoming engaging space than what the other space was. There was a lot of nooks and crannies and there was a lot of kind of darkness. Like, the corridor was really dim and dark and it was not very welcoming. I remember my first time walking into the old place as a case manager, I had a patient in there and you would walk into literally a wall. So, you walked into two double doors, so you walked through an air lock which in itself was confronting, and then you'd be looking

at a wall which had no directions on it. So, you're sort of standing in this dark corridor, and you're already in the clinical space so you didn't know who would be down the corridor, you didn't know where to go. You saw a bit of a light down the end of the corridor so you thought, 'Oh, that must be where I must head.' So, you head down there, and I remember just feeling really overwhelmed and quite scared myself of going into the space. Whereas now you walk in and there's someone around. Like, there's a welcomeness to the unit. There's lightness. You know where you're going to. There's someone to say, 'Oh, hello, this is the office, and this is where you come to.' So, I think that welcoming perspective of knowing where to go when you come into the space is more open now.

Indeed, the increased light was integral to how some of the staff understood the mental health environment; here is how one staff member expressed it:

The things that are really great about it is that the ward is so much lighter. That's, that's the really thing that stands out, like with the courtyards, the windows, it's fresh and it's light and it's bright and, which is much more conducive to mental health, you know, as well.

The value of natural light was also commented on as staff discussed their experiences of the public and private internal corridors. Within the design of a large hospital, it is inevitable that some corridors and some rooms (see subsequent section) will not be able to have access to natural light through windows. How staff represented their own and imagined patient and visitor experiences of a corridor with natural light is illustrative of the value that they gave to daylight. One of the corridors was connected, by way of a large window, to one of the internal courtyards, which filled a particular area of it with natural light. This corridor led to the AAU and OPU, which were the units with the longest walking distance from the reception. To access both units involved walking along the public secure corridor internal to the psychiatric department. The distance and the slow pace of walking of many older patients and visitors gave them more time to attend to the visual and sensory elements of the design of the corridor, including its windows and garden views, and end-of-corridor wall murals. These features were commented on by visitors who appreciated and

Figure 3.3
Daylight in the corridor.
Photo: Melisa Duque.

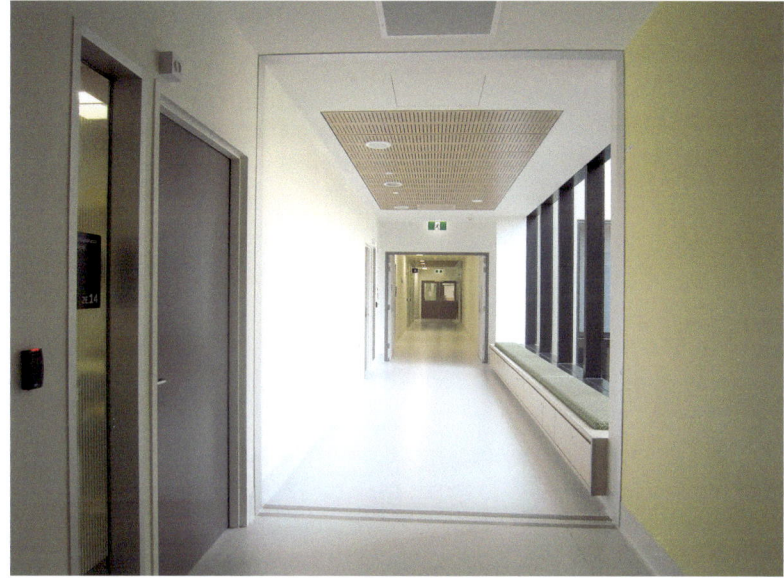

enjoyed the corridors, and the staff often used the walk along the corridor to introduce visitors to the units.

Staff consistently noted and emphasised the value of the natural light they experienced in the corridor with windows:

> You've got a bit more natural light coming in and I think there is some colours or some greenery that you can see.
>
> (Occupational therapist)

> The corridor going down I really like. It's long but it's punctuated by the large courtyard on the left and the seats, and there's the large photo mural at the end and there's a bit of stuff on the ceiling halfway down which creates points of interest as you go along which break down the journey and it doesn't seem like you're going down a long white tunnel.
>
> (Consultant)

> The patient corridor if you like, the more public corridor, I think it's nicer because it's a bit more open, glass, so there's more light.
>
> (Consultant)

This example, on the one hand, adds to and endorses the argument advanced in health environment research, outlined earlier, which associates access to natural light with improved well-being. However, as we show in the next section, while different types of lighting certainly formed part of the experiences and narratives through which corridors were valued, light cannot be seen as determining how people feel. Rather, we suggest that how light participates in people's feelings is contingent on values.

Light and values

In this section, we account for two other examples, which endorse the argument that access to daylight participates in constituting circumstances for the generation of well-being. Yet they also show how, while light is valued in terms of its relationship to well-being, the value attributed to light is always contingent on professional values.

In a large hospital building, it is inevitable that there will be rooms with no natural light. In the spatial design of the Bendigo Hospital, patient access to natural light had been prioritised, meaning that most staff offices had no external windows. However, given the importance staff put on daylight in the patient spaces, we were surprised that they described the artificial lighting system in staff offices so positively. For example, this is how one registrar expressed it: 'Very light. I like that you can adjust the light as well, if you hold it you can dim it or make it brighter'. Sometimes staff struggled with the lack of daylight and fresh air in the offices yet nevertheless praised the offices; here is how one participant put it: 'In here it's a little bit stuffy, and I hate that there's no windows, so we can't see nature. But the space otherwise is pretty good' (Psychologist). Indeed, although staff pointed out that while they wished for windows, they believed that patients should be prioritised in terms of access to fresh air and daylight. This gave precedence to values that were associated with a culture of care and suggested that sacrificing light for others can be a mode of caring that is entangled with professional values.

A further example relates to how staff understood their own needs in the context of a stressful and demanding workplace. Elsewhere, we have written in detail about the requirements that staff across two of the psychiatric units had for staff break areas and outlined how other existing literature emphasises the importance of views of nature and daylight in the areas where hospital staff take breaks (Pink et al. 2020). However, our research showed that here, likewise, specific professional values and needs tended to lead staff to prioritise different elements. Here, priorities of the staff in the most demanding of the psychiatric units were not to access daylight or views of the outdoors but rather to find a quiet, secluded space where they could unwind and debrief without needing to be social unless they wished to. While some staff took breaks outdoors or in the light-filled atrium, more often than not, staff wished to take breaks in quiet, private spaces. Therefore, often, they sought out empty internal meeting rooms, medication rooms and other spaces with no natural light as spaces of refuge, in which they could recover their sense of well-being in solitude or in debriefing with a close colleague (Pink et al. 2020).

In a large hospital building, access to daylight is never an evenly distributed resource, and moreover, it is shared with others in relation to a particular set of values related to care. Staff found their own lack of access to daylight to be inevitable and, indeed, prioritised different experiences within the building design in seeking to ensure their own well-being. Thus, access to daylight does not have a direct and objective impact on staff well-being, rather it is bound up in the values that pertain to their professional culture. Its impact is always relative and contingent. This is not to advocate that staff would not benefit from naturally lit offices and break rooms; indeed, the evidence suggests that they would. Rather, it reveals how the circumstances for well-being are contingent on a number of factors, and their relative importance cannot be known until research is undertaken with staff in the everyday realities in which they work.

Values supporting value

In this chapter, we have outlined a series of different ways in which light can be associated with value and values. In existing research, we have shown that the experience of light in patient spaces has been treated as a resource that can be audited and is shown to bring about measurable benefits – that is, it can be seen to have value as a commodity. However, in contrast, the conviction of the staff that patients should be prioritised over them in allocations of natural light is not a commodity, or a measurable or reportable element of building design. It is created by their unwritten professional values. Thus, Graeber's distinction between value and values is manifested in the way light becomes meaningful, whereby there is an interdependency between the visible value attributed to the resource of natural light and the invisible values of staff in their belief that patients should be prioritised above them in the allocation of access to natural light. Indeed, as Graeber puts it of values, 'they will always resist formalization – to even suggest doing so is at best odd, and probably offensive' (2013: 224).

The examples discussed earlier reveal that the experience of light is both a valued resource and also imbued with professional values. These qualitative ways in which value and values are associated with light, as we stressed at the beginning of this article, complicate but are not incompatible with quantified measures of well-being. For instance, in the OPU, the increase in natural light was understood as an improvement. As one nurse put it, 'Natural light has been a positive in many ways. In the old ward, we had to turn on the light even in the mornings' (Nurse). The head nurse in OPU also suggested that the fact that the number of falls in the unit had decreased could be attributed to the unit being lighter, as she explained this to us:

> Recently we looked at our falls data in comparison to the old unit and here . . . And we have significantly reduced, so from 2016 we were 358; last year 120. So, in the falls meeting that I attend we were asked, you know, why? How has this happened? So I guess the factors that I've come up with is this unit is much lighter and brighter. We don't have speckles on our floor, like the old unit used to have in the vinyl. Patients used to bend down to try and pick them up. And the other factor is that a lot of our falls risk patients, you know the higher acuity, the more distressed and agitated, a nurse in our high dependency unit where our staffing ratio is higher. So they're giving a much higher observation. So that was really interesting. That was really exciting for us to, that's a significant reduction.
>
> (OPU nurse)

Conclusion

As we have outlined earlier, existing research into the ways that staff and patients experience lighting in hospitals tells us that lighting is beneficial to well-being. When we add to this further knowledge about why or how the experience of light is inflected with specific social, cultural and professional dimensions, we can better understand the implications of lighting design for well-being. As we have mapped out, a phenomenological design anthropology of lighting offers a methodology that can make a difference in this field by bringing to the fore the contingent circumstances and specific workplace values and experiences that participate in determining when and where natural and artificial light and lighting are most valued.

Such a methodology is inevitably always site specific since different professional and local cultures may bring different values and generate different experiences. However, our methodology is precisely designed to account for this variation. That is, our anthropological theories of light are abstract, rather than specific; they are modified when tested out in the specificity of life as lived and experienced. Our contribution to anthropology is in this mode of theory building. Here, we suggest that a design anthropological theory of light, which accounts for the contingent circumstances and modes of improvisation through which experiences of light emerge needs to be accounted for as we seek to understand light itself as an experientially emergent phenomenon. Our contribution to design is to propose that attention to these contingent, locally and professionally specific circumstances can offer significant insights to the design process. Such knowledge adds to existing standards that demonstrate the well-being benefits of natural light through quantitative studies, to show how these benefits might come about in everyday life circumstances and practice.

Note

1. In a 2018 keynote lecture to the WWNA – 'For Human Centric Design: Work With Human Centric Specialists!' (https://www.applied-anthropology.com/video/) – Jose Manuel dos Santos, an industrial designer and head of design and user experience at Signify (Americas), acknowledged the value of anthropology in lighting design.

References

Akama, Y., Pink, S., & Sumartojo, S. (2018). *Uncertainty and Possibility: New Approaches to Future Making*. London: Bloomsbury.

Andrade, C., Lima, M. L., Fornara, F., & Bonaiuto, M. (2012). Users' Views of Hospital Environmental Quality: Validation of the Perceived Hospital Environment

Quality Indicators (PHEQIs). *Journal of Environmental Psychology*, *32*(2), 97–111.

Annemans, M., Stam, L., Coenen, J., & Heylighen, A. (2017). Informing Hospital Design through Research on Patient Experience. *The Design Journal*, *20*. https://doi.org/10.1080/14606925.2017.1352753 Open Access.

Bates, V. (2018). "Humanizing" Healthcare Environments: Architecture, Art and Design in Modern Hospitals. *Design for Health*, *2*(1), 5–19.

Bille, M. (2019). *Homely Atmospheres and Lighting Technologies in Denmark: Living with Light*. London: Bloomsbury.

Boyce, P. (2014). *Human Factors in Lighting*, 3rd Edition. London: CRC Press.

Buse, C., Martin, D., & Nettleton, S. J. (2018). Conceptualising 'Materialities of Care': Making Visible Mundane Material Culture in Health and Social Care Contexts. *Sociology of Health and Illness*, *40*(2), 243–255.

Dalke, H., Little, J., Niemann, E., Camgoz, N., Steadman, G., Hill, S., & Stott, L. (2006). Colour and Lighting in Hospital Design. *Optics & Laser Technology*, *38*(4–6), 343–365.

Dijkstra, K., Pieterse, M., & Pruyn, A. (2006). Physical Environmental Stimuli That Turn Healthcare Facilities into Healing Environments Through Psychologically Mediated Effects: Systematic Review. *Journal of Advanced Nursing*, *56*(2), 166–181.

Edensor, T. (2012). Illuminated Atmospheres: Anticipating and Reproducing the Flow of Affective Experience in Blackpool. *Environment and Planning D: Society and Space*, *30*, 1103–1122.

Graeber, D. (2013). It Is Value That Brings Universes into Being. *HAU: Journal of Ethnographic Theory*, *3*(2), 219–243.

Huisman, E. R. C. M., Morales, E., van Hoof, J., & Kort, H. S. M. (2012). Healing Environment: A Review of the Impact of Physical Environmental Factors on Users. *Building and Environment*, *58*, 70–80.

Ingold, T. (2010). Footprints Through the Weather-World: Walking, Breathing, Knowing. *Journal of the Royal Anthropological Institute*, *16*, S121–S139.

Ingold, T. (2011). *The Perception of the Environment: Essays on Livelihood, Dwelling and Skill*. London: Routledge.

Irving, A. (2017). The Art of Turning Left and Right. In J. Salazar, S. Pink, A. Irving, & J. Sjoberg (Eds.), *Future Anthropologies*. Oxford: Bloomsbury.

Iyendo, T. O., Uwajeh, P. C., & Ikenna, E. S. (2016). The Therapeutic Impacts of Environmental Design Interventions on Wellness in Clinical Settings: A Narrative Review. *Complementary Therapies in Clinical Practice*, *24*, 174–188.

Jensen, R., Strengers, Y., Raptis, D., Nicholls, L., Kjeldskov, J., & Skov, M. (2018). Exploring Hygge as a Desirable Design Vision for the Sustainable Smart Home. DIS'18, 9–13 June 2018, Hong Kong. https://doi.org/10.1145/3196709.3196804

Knight, C. (2010). Field Surveys of the Effect of Lamp Spectrum on the Perception of Safety and Comfort at Night. *Lighting Research and Technology*, *42*(3), 313–329.

Martin, D., Nettleton, S., Buse, C., Prior, L., & Twigg, J. (2015). Architecture and Health Care: A Place for Sociology. *Sociology of Health & Illness*, *37*(7), 1007–1022.

Nejati, A., Shepley, M., Rodiek, S., Lee, C., & Varni, J. (2016). Restorative Design Features for Hospital Staff Break Areas: A Multi-Method Study. *HERD: Health Environments Research & Design Journal*, *9*(2), 16–35.

Pierri, P. (2018). Participatory Design Practices in Mental Health in the UK: Rebutting the Optimism. *Design Issues*, *34*(4), 25–36.

Pink, S. (2004). *Home Truths*. Oxford: Berg.

Pink, S. (2012). *Situating Everyday Life*. London: Sage.

Pink, S. (2015). *Doing Sensory Ethnography*. London: Sage.

Pink, S., Duque, M., Sumartojo, S., & Vaughan, L. (2020). Designing for Staff Breaks. *HERD: Health Environments Research & Design Journal*, *13*(2), 243–255.

Pink, S., Fors, V., & Glöss, M. (2018). The Contingent Futures of the Mobile Present: Beyond Automation as Innovation. *Mobilities*, *13*(5), 615–631. https://doi.org/10.1080/17450101.2018.1436672

Pink, S., & Leder Mackley, K. (2016). Moving, Making and Atmosphere: Routines of Home as Sites for Mundane Improvisation. *Mobilities*, *11*(02).

Pink, S., Leder Mackey, K., Morosanu, R., Mitchell, V., & Bhamra, T. (2017). *Making Homes*. London: Bloomsbury.

Pink, S., & Sumartojo, S. (2017). The Lit World: Living with Everyday Urban Automation. *Social and Cultural Geography*. http://dx.doi.org/10.1080/146493 65.2017.1312698.

Shikder, S., Mourshed, M., & Price, A. (2012). Therapeutic Lighting Design for the Elderly: A Review. *Perspect Public Health*, *132*(6), 282–291.

Smith, R. C., & Otto, T. (2016). Cultures of the Future: Emergence and Intervention in Design Anthropology. *Design Anthropological Futures*. eds. Smith, R. C., Vangkilde, T. C., Kjærsgaard, M. G., Otto, T., Halse, J., Binder, T. London: Bloomsbury Academic, 19–36.

Chapter 4

THE MIDWIFERY FEEL OF LIGHT

Stine Louring Nielsen

DOI: 10.4324/9781003182610-4

Stine Louring Nielsen

Lighting in healthcare environments

Lighting in European healthcare environments in general has been designed and implemented to 'meet the needs for visual comfort and performance of people having normal ophthalmic (visual) capacity', as dictated by the DS/EN 12464-1 European standard on lighting of workplaces (2011: 6). Consequently, lighting in conventional healthcare environments is dominated by cool-white lighting as a means to serve and support visual perception for functional and efficient work practices of healthcare staff. Recent developments within lighting technologies and research are, however, expanding the possibilities and potentials of what light can and do in healthcare environments. With the invention and introduction of light-emitting diodes (LED) to our everyday lives within the last decade, the spectral distribution of electrical light is now adjustable like never before. Today, so-called 'white' light is available in different compositions, from a warm amber appearance (2200K) to a more neutral (3000–4000K) or a cool blue-white appearance (6500K) and beyond. This technological development makes it possible not only to imitate the circadian rhythm of daylight but also to tune the spectral distribution of a light source.

In relation to these technological developments, studies are being carried out on the effects of different light spectra in relation to human visual perception and psychophysiology. For example, studies have found that contrast illuminations (such as red and green) may increase visual comfort by supporting visual abilities to detect structures in an environment and lower symptoms of eye strain (Boyce 2014), while circadian lighting (imitating naturalistic daylight and its rhythm) has been shown to support recovery from depression (Golden et al. 2005) and lower pain levels and the intake of pain analgesics (Malenbaum et al. 2008). Additionally, spectral compositions with a warm amber-reddish appearance have been shown to allow relaxation, heighten sleepiness ratings and decrease reaction time (Lockely et al. 2006); while blue-white appearances of light have been shown to suppress the production of melatonin (sleep hormone) and raise cognitive arousal (Wulff-Abramsson et al. 2019) and general attention (Brown 2020).

Informed by this corpus of research, chromatic lighting technologies are now designed and applied in healthcare environments as 'ergonomic lighting' to support visual perception of staff during surgery and as 'circadian lighting' to support physiological circadian rhythms and psychological states of depression and schizophrenia (cf. Chromaviso 2021; LightCare 2021). However, in addition, some healthcare institutions have started installing chromatic lighting designs as part of so-called 'ambient interventions' (WaveCare 2021). This trend especially shows within psychiatric units and maternity wards, broadly motivated by a wish to support a non-medical therapeutic effect of stress reduction and oxytocin release (LaCava 2014). These design interventions are currently being installed worldwide, from Europe to China and North America. In Denmark alone, within the last 7 years, 93 'sensory rooms' have been installed within psychiatry and 29 so-called 'sensory delivery rooms' have been installed in maternity wards in 3 out of 5 regions (WaveCare 2021). These rooms incorporate a mix of blurred nature images, sound and chromatic lighting – consisting of bluish, amber and reddish hues – in addition to the conventional white illumination of healthcare environments (ibid.; see Figure 4.1).

The light sources of the sensory delivery rooms include LED spots and a dynamic soft screen controlled by a tablet, viewing four preset guiding programs. These programs are for midwives to apply and adjust during the different phases of birthing – from 'Arrival' and 'Relaxation' to 'Breathing' and 'Welcome'(ing) of the child. As shown in Figure 4.1, the 'Breathing' program imbues a cold bluish light and the others warmer reddish/orangey hues – in addition to

Figure 4.1
Chromatic illumination settings in a sensory delivery room.
Image: Stine Louring Nielsen.

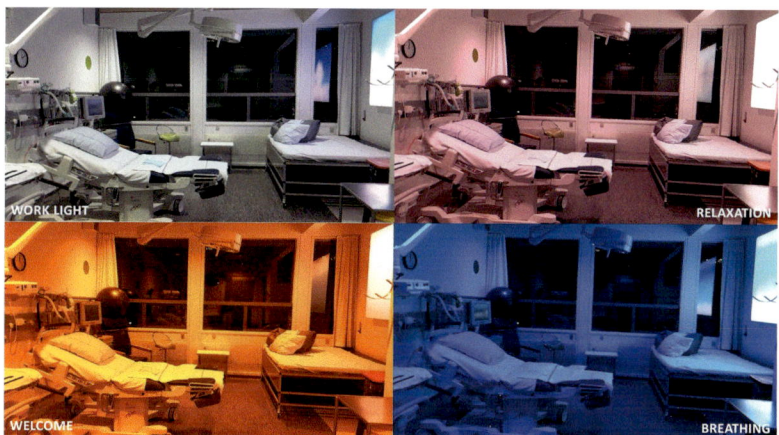

the conventional white 'Work Light' in delivery rooms and healthcare environments in general.

However, as I have also argued elsewhere (Nielsen et al. 2020), midwives working in sensory delivery rooms in Denmark point towards a discrepancy between the current regulatory requirements and the designerly intentions of these preset programs and their actual applications of lighting during the process of delivery. In the following discussion, I unfold this argument by firstly laying out a conceptual framework for understanding the human body as more than psychophysiological and visually perceiving – as put forth by the current regulatory requirements of lighting in healthcare environments and the designerly intentions of the 'ambient intervention' in delivery rooms. By means of two prominent thinkers on atmosphere, I argue for an expanded understanding of the body as felt and feeling within the context of lit space. To embody this framework of thinking, I move on to present a selection of empirical data on how midwives practice lighting in conventional and sensory delivery rooms using their own sensory awareness and attention towards the body as felt and feeling and its attunement to its surrounding lit environment. Lastly, I turn to a discussion of how this sensory awareness and attention emphasises lighting as an aesthetic design element with potential to affect the human body in multiple ways within and beyond current regulatory requirements and designerly intentions of lighting in healthcare environments.

The atmospheric matter of body and space

To initially expand conventional understandings of the human body within the context of lit space, one may firstly attempt to approach space as more than a solely geometrical entity and the body as more than psychophysiological and visually perceiving. As a means to this, I turn to philosopher and phenomenologist Gernot Böhme, who discusses the interstices of the sensory body and affective spaces from a concept of atmosphere. Essentially, Böhme argues that light has the power to tint spaces with a certain feel. On the matter of chromatic lighting, he more specifically states how, 'with the aid of illumination, entire scenes can be overlaid with a colour-modifying hue, lending a characteristic mood to the whole' (2017: 203). In general, he notes how illumination affects the presence and enhances the feel of space by enabling the ability of objects to step out of their own tangible boundaries and impose themselves on the environment by ecstatic means (Böhme 1993, 2013b). Within this framework, Böhme, thus, fundamentally argues that light serves as a generator of atmosphere by pointing out how illuminations imbue a certain 'spatially extended quality of feeling' (1993: 118).

Understanding light as an enabler of allowing spaces to exceed their physical boundaries by atmospheric means, Böhme points to how a classical geometric understanding of lit space is insufficient, when concerned with its felt effects. On this matter, he argues how the phenomenon of atmospheres generally positions within two contrasting concepts of space, that is, 'space as a medium of representation' and 'the space of bodily presence' (2003). Where the former concept addresses physical geometrics of space, the latter addresses space from a phenomenological approach to human body-sensory existence (ibid.). From this conceptual distinction of space, Böhme argues how spaces occupied with sensing bodies become qualitative spaces of atmospheric affect. Along these lines, Böhme fundamentally defines atmospheres as 'manifestations of the co-presence of subject and object' (Böhme 1998: 114), as roughly represented in Figure 4.2.

Essentially, Böhme, thus, emphasises how, without any present and perceiving body, atmospheres cannot exist. In doing so, he moreover argues how 'the spatial shape of architecture is not merely a matter of what you see, but is rather experienced in and by the body, as if it were realised internally' (Böhme 2013a: 21). In continuation to this argument, Böhme introduces yet two contrasting concepts – that of 'the body' and 'the mindful body' (ibid.: 26–31). Where the former addresses the body as an objective entity between other bodies in geometric space, the latter addresses the sensitive body in affective space. In this regard, Böhme's concepts show relevant for lighting designers because they point towards the effects of lighting beyond its physical ability to light up a geometrical space for us to visually perceive. Following Böhme, lighting becomes an aesthetic design element of atmospheric potential to attune our bodily state of being in space by affecting 'the sense or feeling one has of the space where one is' (ibid.: 27). In other words, lighting becomes a sense of spatiality.

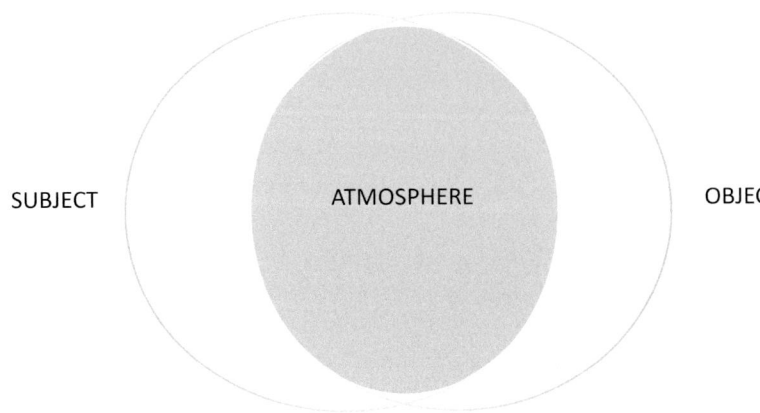

Figure 4.2
Böhme's concept
of atmospheres, as
manifestations of the
co-presence of subject
and object.
*Image: Stine Louring
Nielsen.*

However, Böhme omits to elaborate much or dwell on the characteristics of our sense of spatiality or 'bodily states of being' in lit space (1993: 122). Instead, he may be said to attend more to the presence of objects than that of the subjective body by largely focusing his conceptualisations of atmosphere around design elements of architecture, building materials, light and sound (2002, 2013b, 2017). As a means to address the bodily sensation of being in lit space, another prominent theorist of atmosphere, Hermann Schmitz, helps us to address the shortcomings of Böhme's work. Fundamentally, Schmitz does so by centring his conception of atmosphere around the human body as a felt sphere of affective involvement, conceptualized as *the felt body* (Schmitz 2017, 2009). In this regard, atmospheres, according to Schmitz, precedes the splitting of subject and objects by being both experienced – and inseparable from the felt body (Schmitz et al. 2011: 246). Hence, the felt body is inextricable from an understanding of space. It is in itself a spatial phenomenon – a mode of being spatial, of taking space, of being in and of a space – an elementary spatial dynamic (Slaby 2019: 278), as visually represented in Figure 4.3.

Within Schmitz's conceptual framework, a person's felt body is further understood to be what a person can feel or sense of himself/herself in the sphere of his/her material body, without falling back on the five senses or on habituated ideas of the body (Schmitz 2016: 3; 2009: 15–16). According to Schmitz, the felt body includes all 'corporal stirrings', from *inner* 'pure stirrings', such as itching, and 'emotional stirrings', like joy, to *outer* 'perceived movement', such as

Figure 4.3
Schmitz's concept of atmosphere, as experienced from and part of the human felt body.
Image: Stine Louring Nielsen.

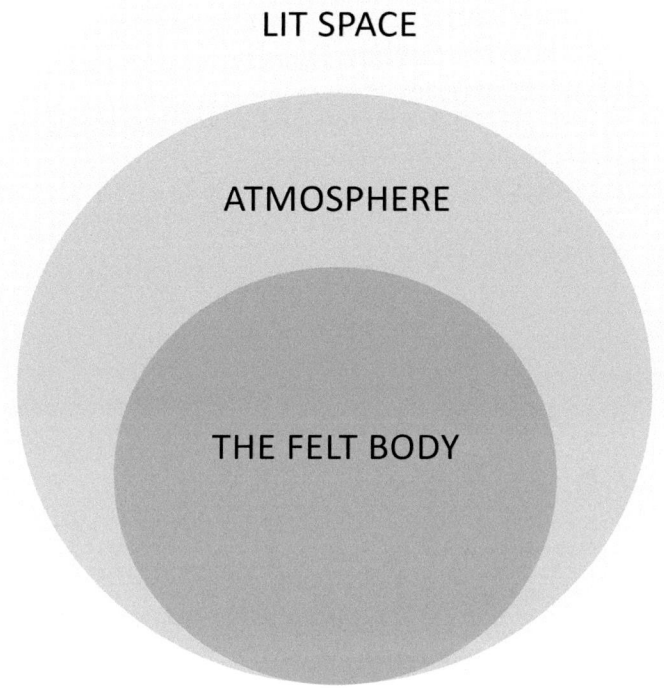

LIT SPACE

ATMOSPHERE

THE FELT BODY

jumping and 'irreversible corporal directions', like gazing and breathing out (2016: 3). As such, Schmitz's conceptualisation of the felt body as a spatial phenomenon points to how lighting may ultimately attune the body in space in inner bodily sensations and outer movements.

In conjunction, Böhme's and Schmitz's phenomenological conceptualisations of atmosphere bring forth an understanding of the human body as felt and feeling within the empirical framework of lit space – from delivery rooms and beyond. Within this framework, bodies and lit spaces are more than physical entities. Rather, they are attuning and attuned by atmospheric means. As such, the conceptualisations of atmosphere and the felt body enable an understanding of lighting as something more than a design element affecting our visual perception of space and our psychophysiological body of circadian rhythms and moods. Instead, it enables an understanding of lighting as an aesthetic design element and a constitutive generator of atmosphere, with potential to attune bodily sensations and movements in space. In this regard, how we as human (felt) bodies practice lighting is as much a matter of how it feels as it is a question of what might be deemed as objectively 'optimal' for particular tasks or uses of lighting. Accordingly, attending to the atmospheric potential and effects of light can assist lighting designers in working towards meeting sensory needs of the human body in lit space beyond current standards on lighting. This attention offers a perspective that attends to and qualitatively values how lighting is practiced and experienced starting from a sensory awareness, rather than using conventional forms of measurement that can miss vital perspectives of people, such as midwives, who work closely and expertly with customisable lighting.

To exemplify this, I now turn to the particular case of how midwives at a Danish maternity ward practice conventional and chromatic lighting during the process of birth. The empirical data of this case was collected during a four-month ethnographic field study at Rigshospitalet's maternity ward in Copenhagen; Rigshospitalet is the largest hospital in Denmark, housing 5,500 deliveries per year. Their maternity ward includes both conventional and sensory delivery rooms, in which I carried out approximately 700 hours of participant observations during 62 processes of birth and conducted 30 formal and countless informal interviews with midwives, in addition to mapping out and documenting their practices in videos and photos. Essentially, the study was carried out by applying a sensory ethnographic approach (Pink 2015), where I explored both the cultural judgment of sensory inputs as well as understanding midwifery practices of facilitating birth through a sensory lens – with an essential orientation towards midwives' sensory awareness of lighting. As I shall lay out, the midwifery lighting practices at Rigshospitalet reveal their particular feel of how lighting holds the potential to attune human bodies in the delivery room by atmospheric means, when studied from a sensory ethnographic approach and viewed through the conceptual framework of atmosphere and the felt body.

(Light) setting the (birthing) scene

'I rarely use ceiling lights inside these delivery rooms. I find it annoying that it has to be a part of it all. Ceiling lighting is such an institutional thing', the midwife Sarah explained to me during one of our night shifts together. As such, she pointed towards a common cultural judgment and feel of the conventional cool-white lighting among midwives working in the delivery rooms of Rigshospitalet. That is, in general, this mode of lighting was considered as 'an institutional thing', imbuing an imperious feel. This was particularly articulated by midwives characterising the lighting as 'overbearing' and 'noisy' in terms of disturbing their peace of mind. For one midwife, the annoyance and noisiness were experienced to the extent that she claimed how 'the worst thing is to take over a delivery room where all the lights are on', while another midwife described how 'I am not as relaxed if there is full flare on the electricity in there'.

As a means to counteract the 'noisy' and imperious feel of traditional hospital lighting, midwives would perform several lighting practices in the delivery room. This, for example, showed when midwives would set the scene for birthing by tuning the lighting in the delivery room from the very beginning of each birth process – even before welcoming the birthing mother and her partner(s). This was, for example, the case when I joined the midwife Alice during one of her evening shifts. We had just been assigned to assist a birthing mother, but before guiding her into one of the conventional delivery rooms, Alice walked to the room for an initial lighting intervention. 'What are you doing with the lights?', I asked her. Alice replied:

> I'm trying to get it as muted as possible, to create a sort of cave-like atmosphere. So that there isn't this spotlight shining down, unless we need it in an emergency situation or need to look properly at the child. Then we turn it up properly. But otherwise I try to turn off the top lights as much as possible and only have on the lights above the cabinets. Some of the rooms have a small architect lamp you can just switch on instead. Those I like.

Alice's lighting intervention exemplified a shared practice among the midwives to turn off or dim the lighting in the conventional delivery rooms as a means to support 'a sort of cave-like atmosphere'. Hence, instead of flooding space with an even light distribution by applying spots and ceiling lights, midwives generally favoured turning on small lights located under the cabinets and creating spaces of focal glow by, for example, pointing the operation and architect lamp to the wall, as shown in Figure 4.4.

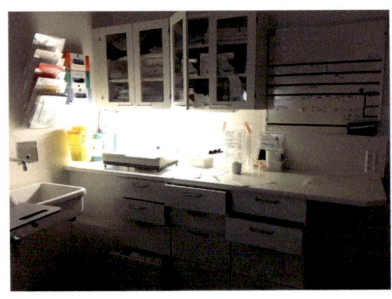

Figure 4.4
Midwifery lighting interventions in conventional delivery rooms.
Photos: Stine Louring Nielsen.

As such, midwives essentially showed their concern with zoning and muting sensory stimuli in the delivery room in consideration to the location of the birthing mother. This sensory attention towards the felt body did not only apply for the lighting. Midwives would also care for the sensuous aspects of a birth by, for example, muting the alarm system when entering the delivery room, throwing away plastic bags of vomit to eliminate the stench, covering up visual bloodstains from bed frames and floors, or leaving drawers and cabinets open to avoid the noise of slamming during the process of delivery (see Figure 4.4). As such, the midwifery practices of setting the scene and tuning the lighting in the delivery rooms of Rigshospitalet point to an essential aspect of being a midwife, namely, the aspect of supporting intimacies. Anthropologist Mikkel Bille, midwife Anne Barfoed and I have argued this point elsewhere (Nielsen et al. 2020), which moreover corresponds with the social history and ancient rise of midwifery, rooted in solidary aid and trust (Böhme 1984). Emphasising this rootedness further in relation to lighting, some midwives at Rigshospitalet would, for example, run around the ward searching for an architect lamp just to avoid having to turn on the ceiling lights in a conventional delivery room. The sensory attention towards supporting atmospheres of intimacy was sometimes extremely explicit in practice.

The predominant midwifery feeling towards traditional hospital lighting more generally corresponds with a common articulation among other visitors of hospitals who have defined its lighting as a white 'killer-light', 'flat' and like 'infinite and everywhere' (Stenslund 2017), in addition to more general articulations of bodily sensations in bright-white lighting as generating 'a higher tone in my body', a feeling of being more 'in my mind and not my body' or the experience of being able to 'cut through things' (Nielsen et al. 2021). Relatedly, one midwife commented prior to adjusting the lighting in a conventional delivery room: 'Just try and feel it, it's not very enjoyable, this light we are sitting in right now . . . I just don't like being in it. I would never choose this light if I were to feel safe or relaxed or work'.

(At)tuning lighting during the process of delivery

For the midwives, it all came down to trying to create a 'home-like, cosy feeling' in the delivery room, where the couple did not feel insecure about 'what is going to happen here?!' when entering. In this regard, by the introduction of chromatic lighting into the delivery room, light setting the birthing scene became a matter of more than turning off and dimming lights for the midwives at Rigshospitalet. In the sensory delivery room, their midwifery lighting practices expanded to a matter of also tuning and attuning to saturated spectral compositions of lighting throughout the birthing process.

This, for example, showed during an evening shift I shared with the midwife Sarah. She and I had just assisted the delivery of a healthy baby boy in the sensory delivery room. Over the course of five and a half hours, Sarah had been supporting and easing the coming mother by resting slightly stroking hands on her body and speaking appraising comments, like 'you are doing great!' and 'don't worry, nobody looks good in those', referring to the hospital net underpants she was wearing. Meanwhile, I had fetched cold lemonade and heating pads, along with writing down numeric values on the whiteboard. Throughout the birthing process, Sarah had carried out midwifery practices of keeping track of numerous technologies of CTG scans and IV poles, using blood pressure devices and applying 'bee sting' injections for pain release. In addition, she had adjusted the bluish, amber and reddish hues of illumination via the tablet mounted by the entrance. By the end of our shift, Sarah reflected on her lighting practice during the birthing process:

> I changed the setting because I thought we just needed a change of energy. And then I changed it again because I thought we needed another energy change. A little more like a cave and down in . . . not in pace, because we still had to get the contractions going,

but that the contractions would be better stimulated by a calm and cheerful condition. And finally, it was a combo, the last change, that 'Welcome thing', it was a combination of calm and enclosing . . . you know that womb-like feeling and a welcoming to the child . . . For me it seems to create safety and cosiness. And intimacy.

By adjusting the colours of illumination installed in the sensory delivery room, Sarah, according to herself, supported a 'change of energy'. From seeking to calm in the beginning, to then 'boost some energy' into the situation and, lastly, to support an enclosing 'womb-like feeling', she articulated a practice of lighting in the delivery room based on a sensory awareness of how different hues and intensities of illumination attuned certain bodily sensations and movements.

Essentially, Sarah noticed how 'I always think I'm trying to work with the lighting in the delivery rooms. And here it was possible to make it a little better . . . because it was coloured'. As such, she pointed towards a common experience of the sensory delivery room providing an additional midwifery tool to attune the birthing process by replacing the 'flat' and 'noisy' white lighting of the conventional delivery room with saturated hues of light. In practice, this meant that midwives would rarely turn off the chromatic lighting in the sensory delivery room but leave it on throughout the process of delivery. This, for example, showed when midwives kept the lighting on while keeping track of CTG scans on computer screens or while stitching up the perineum after childbirth, as shown in Figure 4.5.

In relation to midwives leaving the chromatic lighting on in the sensory delivery room during the different phases of birth, they expressed a sensory awareness of how the lighting generally supported a particular sphere of intimacy. For example, one midwife described how 'Something else just happens . . . you get wrapped

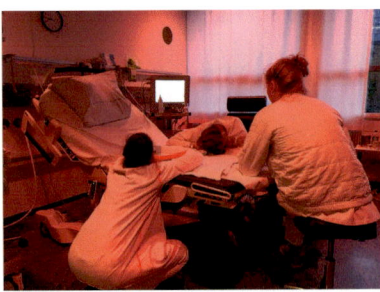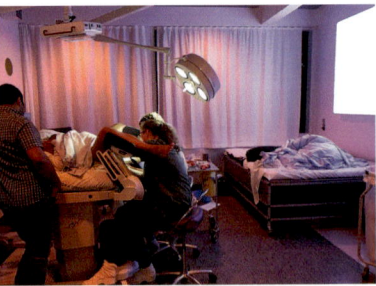

Figure 4.5
Midwifery practices in chromatic lighting during the process of delivery.
Photo: Stine Louring Nielsen.

in your own bell jar in some way and forget about time and space . . . I think a different ease sets in here'. This sensory experience of a certain 'ease' setting in the sensory delivery room was not only noticed by midwives working at Rigshospitalet. Other midwives working these rooms expressed feeling it too. Hence, in an evaluation of the first sensory delivery room in Denmark, midwives are quoted to experience 'a certain atmosphere' and 'quietness' within these spaces (Jensen 2015). One midwife even expressed her bodily sensation of the space by stating, 'I would almost say that one could describe it a bit like having a velvet glove underneath' (ibid.: 4).

These bodily sensations of feeling 'quietness', being at 'ease' and 'having a velvet glove underneath' did, however, not only manifest in midwives' experiences of inner bodily sensations. They also manifested in outer bodily movements and gestures of midwives, doctors and assistants working in the sensory delivery room at Rigshospitalet. For example, doctors and assistants entering the sensory delivery room during the first minute would acclimatise their voices and drop from a relatively high and fast level to a lower and slower one. In accordance with this, one midwife described how 'I actually think that I am a bit more careful in there . . . I think that I am a little slower'. While another one revealed how the lighting had more general implications for the bodily movements in the sensory delivery room by stating how, 'with pink lighting and ocean waves, you can hardly have your shoulders up anywhere. At least, then you really have to want it a lot'. As such, a calm, steady feeling was experienced as purposefully supported by the lighting interventions of midwives at Rigshospitalet, spurred from their particular sensory awareness and attention towards the felt bodies in the delivery room.

The felt body in midwifery lighting practices

The bodily sensations, experiences and practices of lighting articulated by midwives at Rigshospitalet ultimately highlight how lighting serves as a midwifery tool to support the process of birth inside the delivery room, in conjunction with numerous medical technologies and professional skills. In this regard, for the midwives studied, the particular mode of lighting served as more than a tool for just lighting up the physical space of the delivery room to 'meet the needs for visual comfort and performance of people having normal ophthalmic (visual) capacity', as dictated by the European standard on lighting (2011: 6) and the designerly intentions to support a non-medical psychophysiological response of oxytocin release (LaCava 2014). Rather, the lighting served as a generator of atmosphere by imbuing a certain 'spatially extended quality of feeling' (cf. Böhme 1993: 118). From this sensory awareness of the attuning potential of lighting, midwives generally showed to counteract the traditional white lighting in delivery rooms by dimming the lighting in conventional delivery rooms and applying the orangy and reddish lighting in the sensory delivery room. For according to them, these modes of lighting intensity and hue supported the attunements of 'a sort of cave-like atmosphere', a 'home-like, cosy feeling' and sensations of 'having a velvet glove underneath'.

This sensory awareness of the potentials of dimmed and warmer lighting shared among midwives at Rigshospitalet largely corresponds with greater cultural practices of Danes, who are notorious for their using of candlelights for orchestrating sensations of intimacy and 'secureness' at home (Bille 2019), and more specific studies of bodily sensations in warm illuminations, pointing towards predominant sensory experiences of 'being in an onion', feeling 'like a mother carrying you' or being 'pushed down' (Nielsen et al. 2018; Nielsen et al. 2021). Yet besides revealing a general sensory awareness of lighting as a generator of atmosphere with potential to attune inner bodily sensations and stirrings of emotions, the midwifery lighting practices at Rigshospitalet also revealed a sensory awareness to how warmer hues of lighting served to affect apparent movements of the body, such as lowering shoulders, voices and pace in the delivery room, and how this helped midwives and mothers alike in the labour of birthing.

Ultimately, the midwifery lighting practices carried out at Rigshospitalet reveal a certain sensory awareness among midwives of how different modes of illumination may attune certain bodily sensations and movements in the delivery room, from their initial setting of the birthing scene via dimming lighting interventions to their 'changing of energies' during the process of birth. From this awareness, midwives practiced lighting from an overarching attention towards supporting the felt body in the delivery room by atmospheric means, rather than favouring the conventional white lighting in healthcare environments to support visual perception for functional and efficient work practices of healthcare staff or directly following designerly intentions to support a non-medical psychophysiological response of oxytocin release.

Implications for lighting design

Consequently, the sensory attention towards the felt body identified among midwives at Rigshospitalet points towards a discrepancy between the regulatory requirements and designerly intentions of conventional and chromatic lighting in healthcare environments and their actual application in the context of the delivery room. This discrepancy seems to rest on different ontological understandings of the body as predominantly visually perceiving and psychophysiological on the one hand or more expansively felt within the empirical context of healthcare environments in general on the other. As such, the case of midwives practicing conventional and chromatic lighting in delivery rooms at Rigshospitalet fundamentally points to the importance of considering the entangled multiplicity of the human body when designing lighting.

By addressing how lighting holds the potential to affect the human body beyond visual perception, circadian rhythms and moods, this chapter emphasises how a diverse bodily responsibility ought to be at the forefront of any lighting design, guiding the design of each and every lighting solution within. This is because light is a powerful tool, able to attune spatial visual perceptions and hormone productions but also to affect people's lives in the ways we act and feel, in inner bodily sensations and outer movements. And as such, lighting ultimately serves as a design element working on the body in multiple ways beyond those addressed by current regulatory requirements and designerly trends within healthcare environments.

Putting matters in perspective, atmospheres have the ability to suffuse all spatial contexts, however, in some spaces this is more affectively, emotionally, and sensually profound than in others (Edensor and Sumartojo 2015: 252). This is because atmospheres are essentially generated from a greater unmanageable dynamic force field of multiple elements of people, materials, conditions and constitutions making up our lived world – where lighting plays a role but not necessarily the leading one. As such, the dynamic ontology of atmospheres ultimately points to the conclusion that atmospheres cannot be designed. As Böhme puts it, 'the whole is more than the parts' (1993: 124). Nevertheless, the conditions that give rise to atmospheres can be designed by addressing potential conditions and affordances for the human body to (re)act and interact within, as argued by anthropologist Sarah Pink and geographer Sumartojo (2018). Hence, ultimately, 'atmospheres seems to start, precisely where construction stops' (Wigley 1998).

While we undoubtedly still have much to learn about the effects of lighting on human visual perception and psychophysiology, we might have even more to learn about how particular modes of lighting attune human bodily states of being in lit space and the implications of this for both human activity and lighting design. With new perceptual pleasures now available to people via the technical mastery of light and adjustments to intensities and spectral compositions of chromaticity, these learnings are more crucial than ever. If we want to design for the human body living within lit space, regulatory requirements and designers of lighting must now attend to how we as humans actually feel and sense our body within our new tinted world.

References

Bille, Mikkel. 2019. *Homely Atmospheres and Energy Technologies in Denmark. Living with Light*, 1st ed. London: Bloomsbury.

Böhme, Gernot. 1984. "Midwifery Science: An Essay on the Relation between Scientific and Everyday Knowledge." In *Society & Knowledge – Contemporary Perspectives in the Sociology of Knowledge & Science*, edited by Nico Stehr and Volker Meja, 2nd ed., 373–392. New Brunswick and London: Transaction Publisher.

Böhme, Gernot. 1993. "Atmosphere as the Fundamental Concept of a New Aesthetics." *Thesis Eleven* 36 (1): 113–126. https://doi.org/10.1177/072551369303600107.

Böhme, Gernot. 1998. "Atmosphere as an Aesthetic Concept." *Daidalos – Berlin Architectural Journal* 68: 112–115.

Böhme, Gernot. 2002. "Atmosphere as the Subject Matter of Architecture." In *Natural History*, edited by Philip Ursprung, 1st ed., 398–406. Montreal: Canadian Centre for Architecture & Lars Müller Publishers.

Böhme, Gernot. 2003. "The Space of Bodily Presence and Space as a Medium of Representation." *Transforming Spaces: The Topological Turn in Technology Studies* 2: 1–7. http://www.ifs.tu-darmstadt.de/gradkoll/Publikationen/space-folder/pdf/Boehme.pdf

Böhme, Gernot. 2013a. "Atmosphere as Mindful Physical Presence in Space." *Oase* 91 (Sfeerbouwen/Building Atmosphere): 21–31.

Böhme, Gernot. 2013b. "The Art of the Stage Set as a Paradigm for an Aesthetics of Atmospheres." *Ambiances*, October: 2–8. http://ambiances.revues.org/315.

Böhme, Gernot. 2017. "Seeing Light." In *The Aesthetics of Atmosphere – Ambiances, Atmospheres and Sensory Experiences of Space*, edited by Jean-Paul Thibaud, 1st ed., 193–204. New York: Routledge.

Boyce, Peter. 2014. *Human Factors in Lighting*. Edited by Peter Boyce, 3rd ed. Boca Raton – London – New York: CRC Press – Taylor & Francis Group.

Brown, Timothy M. 2020. "Melanopic Illuminance Defines the Magnitude of Human Circadian Light Responses Under a Wide Range Conditions." *Journal of Pineal Research* 69 (1): 1–14.

Chromaviso. 2021. "Chromaviso – Health Promoting Lighting." https://chromaviso.com.

Dansk Standard. 2011. "DS/EN 12464: Dansk Standard Lys Og Belysning – Belysning Ved Arbejdspladser – Del 1 : Indendørs Arbejdspladser Light and Lighting – Lighting of Work Places – Part 1 : Indoor Work Places."

Golden, Robert N., Bradley N. Gaynes, R. David Ekstrom, Robert M. Hamer, Frederick M. Jacobsen, Trisha Suppes, Katherine L. Wisner, and Charles B. Nemeroff. 2005. "The Efficacy of Light Therapy in the Treatment of Mood Disorders: A Review and Meta-Analysis of the Evidence." *American Journal of Psychiatry* 162 (4): 656–662. https://doi.org/10.1176/appi.ajp.162.4.656.

Jensen, Annemette Lundmark. 2015. "En Kvalitativ Undersøgelse Af de Fødendes Oplevelse Af at Føde På En Sansestue." *Tidsskrift for Jordemødre* 4: 1–7. http://www.jordemoderforeningen.dk/tidsskrift-for-jordemoedre/singlevisning/artikel/en-kvalitativ-undersoegelse-af-de-foedendes-oplevelse-af-at-foede-paa-en-sansestue/

LaCava, Laura. 2014. "Ambient Interventions in Healthcare." Copenhagen. http://www.wavecare.com/uploads/8/0/2/3/8023101/ambientinterventionshealthcare.pdf

LightCare. 2021. "LightCare – The Second Best Light on Earth." https://lightcare.dk/en/

Lockely, Steven, Erin Evans, Frank Scheer, George Brainard, Charles Czeisler, and Daniel Aeschlbach. 2006. "Direct Effects of Light on Alertness, Vigilance, and the Waking Electroencephalogram in Humans." *Sleep* 29 (2): 161–168. https://doi.org/10.5665/sleep.2894

Malenbaum, S., F. J. Keefe, A. C. Williams, R. Ulrich, and T. J. Somers. 2008. "Pain in Its Environmental Context: Implications for Designing Environments to Enhance Pain Control." *Pain* 134: 241.

Nielsen, Stine Louring, Ute Christa Besenecker, Nanna Hasle Bak, and Ellen Kathrine Hansen. 2021. "Beyond Vision: Moving and Feeling in Colour Illuminated Space." *Nordic Journal of Architectural Research* 33 (2).

Nielsen, Stine Louring, Mikkel Bille, and Anne Berlin Barfoed. 2020. "Illuminating Bodily Presences in Midwifery Practices." *Emotion, Space and Society* 37 (November): 9. https://doi.org/10.1016/j.emospa.2020.100720

Nielsen, Stine Louring, Carsten Friberg, and Ellen Kathrine Hansen. 2018. "The Ambience Potential of Coloured Illuminations in Architecture." *Ambiances. International Journal of Sensory Environment, Architecture and Urban Space* (4): 0–27. https://journals.openedition.org/ambiances/1578

Pink, Sarah. 2015. *Doing Sensory Ethnography*. 2nd ed. London: Sage Publications.

Schmitz, Hermann. 2009. *Der Leib, Der Raum Und Die Gefühle*, 2nd ed. Bielefeld und Basel: Sirius.

Schmitz, Hermann. 2016. "Atmospheric Spaces." *Ambiances*, September. https://doi.org/10.4000/ambiances.711

Schmitz, Hermann. 2017. *Kroppen*. Edited by Malthe Strandby Nielsen, 1st ed. Aalborg: Aalborg Universitetsforlag.

Schmitz, Hermann, Rudolf Owen Müllan, and Jan Slaby. 2011. "Emotions Outside the Box – The New Phenomenology of Feeling and Corporeality." *Phenomenology and the Cognitive Sciences* 10 (2): 241–259. https://doi.org/10.1007/s11097-011-9195-1

Slaby, Jan. 2019. "Atmospheres – Schmitz, Massumi and Beyond." In *Music as Atmosphere: Collective Feelings and Affective Sounds*, edited by F. Riedel and J. Torvinen, 1st ed., 274–285. London: Routledge.

Stenslund, Anette. 2017. "The Harsh Smell of Scentless Art: On the Synaesthetic Gesture of Hospital Atmosphere." *Exploring Atmospheres Ethnographically*, 153–171. https://doi.org/10.4324/9781315581613

Sumartojo, Shanti, and Sarah Pink. 2018. *Atmospheres and the Experiential World – Theory and Methods. Atmospheres and the Experiential World*, 1st ed. New York: Routledge. https://doi.org/10.4324/9781315281254-1

WaveCare. 2021. "WaveCare – Ambient Interventions for Health and Wellbeing." https://www.wavecare.com

Wigley, Mark. 1998. "The Architecture of Atmospheres." *Daidalos – Berlin Architectural Journal* 68: 18–27.

Wulff-Abramsson, Andreas, Mads Deibjerg Lind, Stine Louring Nielsen, George Palamas, Luis Emilio Bruni, and Georgios Triantafylidis. 2019. "Experiencing the Light Through Our Skin – An EEG Study of Colored Light on Blindfolded Subjects." *Proceedings – 2019 IEEE 19th International Conference on Bioinformatics and Bioengineering, BIBE 2019*, 609–616. https://doi.org/10.1109/BIBE.2019.00116

Chapter 5

PERCEPTIONS OF SAFETY IN CITIES AFTER DARK

Hoa Yang, Jess Berry and Nicole Kalms

DOI: 10.4324/9781003182610-5

Hoa Yang, Jess Berry and Nicole Kalms

Introduction

In 2018, the United Nations made a projection that two thirds of the world's population will live in urban areas by 2050 (United Nations 2018). As the population expands in urban centres and cities, the demand for the right of freedom to enjoy and use public services and spaces becomes more pertinent. The systems and legislation regarding the right of freedom to enable equal access across all hours of the day and night in urban public space should facilitate such experiences. Large governing bodies, such as the United Nations and UNESCO, have attempted to address inequality with the objective of safe public mobility in the design of cities within global policy frameworks, such as 'Right to the City', which was introduced in 2009 (Brown and Kristiansen 2009).

Despite an awareness of the urban governance and policy changes required to improve safe mobility, research of lived experiences in the past ten years has shown that the problem still persists in many urban centres, particularly in hours of darkness (Goulds et al. 2018; Committee for Sydney 2019). Discrepancies between the lived experiences in demographic factors such as gender, sexual identity, race, age, disability, religion and socio-economic class, have been found to influence urban experiences of safety at night (Aruldoss and Nolas 2019; Wattis Green and Radford 2011). This is consistent with the latest OECD 2020 statistics, where women, on average, feel less safe walking alone at night than men. Similarly in Australia, 1 in 2 women feel unsafe when walking alone at night (Community Council for Australia 2019).

This chapter presents the findings from a collaboration between practicing lighting designers and design researchers. The preliminary findings highlight that there currently exists gendered differences in the night-time experience of Australian cities, and links how current urban lighting design practices could be contributing to feelings of fear and safety for marginalised groups of society. Within this chapter and the wider project, the intent of inquiry was solely concerned with the 'perception of safety' in the night-time experience, as opposed to 'being safe', and related investigations of crime.

This research is contextualised by the significant work undertaken by feminist geographers to understand women's perception of safety in the urban environment (Mehta and Bondi 1999; Pain 1991,

1997; Valentine 1990). We take Grover's (2017: 326) definition that a woman's perception of safety is an 'experiential state where women feel or do not feel safe in a given situation or place' and recognise Pain's (1991: 417) assertion that it is the feeling of safety that is of significance, not whether a space is objectively safe or unsafe but rather 'how these spaces are constructed, [and] what they represent'. In this chapter, we focus on how feelings of unsafety in public space are determined by the interrelated factors of physical space – in this case, lighting – and personal individual experiences. In examining women's perceptions of unsafety in public space, it is understood that women adopt their own approaches to avoid feelings of unsafety, which includes avoiding public space after dark (Starkweather 2007; Beebeejun 2009).

The key aim of the project was to merge two lines of inquiry not often integrated in professional practice – women's lived experiences and their perceptions of safety in lit spaces of the urban environment after dark, and quantitative technical measurements of the urban environment. The mixed-methods framework of seemingly disparate data sets has provided a provisional understanding of how the technical specification of lighting and the design of the urban environment can begin to improve perceptions of safety for women, stepping towards spatial equity to foster more engagement with the urban environment after dark. While preliminary, the findings suggest an urgent revision to current Australian practices in lighting design to incorporate community feedback from marginalised experiences of urban space to inform technical lighting and, urban design specifications that actively facilitate equal access to perceptions safety in urban centres.

The problem: opting out after dark

Lighting design as a specialist discipline for exterior lighting in the Australian context is relatively new, compared with traditional disciplines of architecture, urban design and engineering, with its beginnings in the 1950s (Freedman 1976). Recent night-time development in cities and a desire to invigorate the night-time economy has influenced policies that focus on improving experiences and increasing engagement of the community with the public domain (Committee for Sydney 2019; City of Darebin 2019; City of Melbourne 2020). For such experiences to be facilitated, there is an implicit expectation that the public urban environment will provide equal opportunity for users to feel comfortable and safe across 24 hours of the day. Often, this provision assumes that the transition between daylight and electric light at night is seamless in facilitating the same type of urban experiences. Current research findings, however, demonstrate that the day/night perception of safety is imbalanced in gendered groups, where the presence or absence of light at night appears to play a major factor in affecting gendered and marginalised experiences of safety in urban areas.

In the practice and scholarship of lighting for night-time public spaces, the perception of safety at night is conceptualised as 'reassurance', that is, 'the confidence a pedestrian might gain from road lighting (amongst other factors) to walk along a road [. . .] alone after dark' and that which 'provides the comfort that makes someone feel less worried, less afraid or doubtful and restores confidence' (Fotios et al. 2015: 449). There are many factors that can contribute to an individual's experience of perceived safety on the street. This includes past experiences; sociocultural contexts, such as who is using the street and how; as well as the design of the urban environment (Fotios and Unwin 2013). There are two main types of risks that influence pedestrian reassurance – firstly, the risk of being involved in an accident and, secondly, the risk of being victim to criminal offences, violence or threats. It has been found that the latter type of risk is of most importance for pedestrians in influencing behaviours after dark (Fyhri et al. 2010). This suggests that behaviours after dark are more likely to be curtailed for women, the elderly, people of colour and other marginalised groups for fear of being a victim of criminal offences, violence or threats after dark (Pain 2001).

Women's apprehension of the dangers of public space at night is reinforced by gendered experiences of safety that position women as illegitimate users of the city at night (Andrew 2000). While women's fear of violence and sexual harassment, and sense of vulnerability in public space after dark are consequences of gender inequality and perpetuate the social exclusion of women in these spaces (Menih 2020; Koskela and Pain 2000), policy approaches to women's safety in public space at night place the responsibility on women to adjust their own behaviours. Women's risk-taking behaviour at night is framed implicitly as both bodily and reputational threat (Williams 2008), and the 'safety work' of avoidance and risk minimisation that women undertake is seen as a gendered expectation where women trade freedom of movement for safety (Vera-Gray and Kelly 2020). Unsurprisingly, women are well versed in how to take precautionary safety measures at night, yet this does not mitigate their feelings of unsafety.

The collaborative pilot project discussed in this chapter sought to address the issue of pedestrian reassurance for marginalised people in two ways, as a prototype for useful integration of design research with practice. The first was to incorporate site-specific crowdsourced data from a project that focused on the marginalised lived experiences of gender – the *Free to Be* campaign (2016–2019). This data generated insight into gendered experiences of the urban environment during night and day, providing a starting point to understand perceptions of safety associated with urban places in Melbourne. The second approach was to merge these qualitative experiences with quantitative measurements, to inform an in-depth understanding of the lighting and spatial qualities of the environments in which these negative or positive experiences of safety were reported. Bringing together two discrete sets of data, this study generated a framework with which to consider how the specification of lighting and urban spatial qualities can actively contribute to improve people's lived experiences of the city at night.

The following sections in this chapter foreground a practice-informed methodology of centring lived experiences alongside technical analysis to inform the design of equal access to perceptions of safety at night. The framework challenges current lighting standards and design practice by highlighting the inherent bias in understanding perceptions of safety at night for Australian urban environments.

Evaluation of perceptions of safety at night for pedestrians

In considering the differences that occur within the design of an urban environment between day and night, urban infrastructure remains immobile with one key difference – light. When natural light fades after sunset, the illumination of urban spaces become the domain of electric light, a critical element in facilitating pedestrian reassurance in the experience of night-time. Under the veil of darkness, the transition from natural to electric light and the interplay of light with the urban environment has been found to be one of the most important influencing factors in perceptions of safety (Painter 1996; Fotios et al. 2015, 2019; Peña-García 2015). In the multitude of design decisions that can be made for a new or existing urban site, the specification of urban lighting presents itself as one of easiest to manipulate in improving urban atmospheres at night. As such, the presence and quality of light, the way the specified luminaire is interacting with the spatial qualities in the urban environment and the feelings that are being evoked through these installations are considerations that should be made at the beginning of the design process for any external environment.

Studies into the relevance of urban lighting for reassurance were first theorised and researched in the field of criminology and social psychology. Field studies into the contributing factors of reassurance found that 'darkness' or environments with the absence of light and inadequate lighting in streets are generally perceived as unsafe (Nasar and Fisher 1992; Nasar and Jones 1997). This was later augmented by lighting studies that investigated the topic of pedestrian reassurance in the fields of science and engineering. While studies in criminology suggesting differences in social groups and experience in the presence of brightness has found brightness to improve reassurance for women and the elderly (Painter 1991, 1996), science and engineering studies have largely been preoccupied by an inquiry into the functional relationship between vision and its associated hazards due to light.

These investigations have provided the foundation for the specification of lighting for pedestrian reassurance in the standards (Boyce et al. 2000; Narendran et al. 2016), shaping standard practice and policies for the design of urban infrastructure. We observe, however, that there is missed opportunity in accounting for a wider social understanding of the pedestrian experience. This problematisation of light without constructed human experiences is reflected in the current Australian lighting standards and designs for external environments. The Australian/New Zealand Standard, Lighting for roads and public spaces Part 3.1: Pedestrian area (Caregory P) lighting performance and design requirements (AS/NS1158.3.1:2020) focuses exclusively on the amount of light output and evenness of the spread of light from the light source, which is measured in units of lux and uniformity, respectively. Crucially, the current standards omit the surrounding social and spatial contexts that interact with light to evoke atmospheres of fear, safety and comfort at night.

Qualitative data set: *Free to Be*, Melbourne

Research in design and urban planning has gone someway to considering the personal experience of pedestrians when taking into account safety in night-time urban space. Recognising the need to apply a gendered lens to the understandings of the built environment to address issues of safety, access and agency has resulted in methods, such as 'gender mainstreaming', 'gender-sensitive' design and 'gender inclusive' urban planning (Shaw et al. 2013; Roberts 2018; World Bank Group 2020). Imperative to all these approaches is the inclusion of participatory stakeholder engagement, providing insights into the lived experiences of women and gender-diverse people and the spatial factors that contribute to their perceptions of safety and unsafety in the urban environment. The research outlined here operates within this context. By recognising the value of seeking out and understanding the stories of women, girls and gender-diverse people, designers might ascertain the conditions that impact spatial inequity and reconsider the role of the built environment in perpetuating these inequities. It is important here to note that while technical and design solutions are important, they cannot be divorced from the need to challenge gendered power relations (Koskela and Pain 2000; Beebeejun 2009).

In 2018, lighting design consultants from engineering and design firm Arup in collaboration with a data set generated by Plan International and the Monash University XYX Lab investigated the experiences of young women to analyse the relationship between urban lighting and women's perceptions of safety. The *Free to Be* map described happy/safe places and sad/unsafe places, separated by day and night. The geolocation map asked young women to denote happy or sad, and many annotated with details of incidents and descriptions. Here is an example:

> I pass through here twice a day to get to work and am routinely verbally abused by men. I feel unsafe and would never go through here at night. I wish the police or government would listen to women's stories and do something about this place.
>
> (Age 25, anytime)

The comments and pins were coded to gather commonalities. While the 'happy' or 'sad' designations were primarily formulated in consultation with young women who were asked to identify the areas of the city in which they felt safe or unsafe, the final map was crowd-sourced and gathered perceptions of safety in the city of Melbourne and fed information back to councils, public transport operators and the police. As discussed in the associated report 'Unsafe in the City' (2018), when women referred to 'Lighting', this most often meant 'Poorly lit'. This would often describe places where women thought that others might hide in order to attack them – the fear of dark alleys and footpaths was a common concern. Here's an example:

> Inadequate street lighting in this street and surrounds – very dark and feels unsafe. It means I always walk in the middle of the road after dark, which is putting myself in danger, but this feels safer than the dark footpaths.
>
> (Age 41)

This contrasted with descriptions of safe spaces as being well-lit:

> Love the feeling of walking towards the arts centre and gallery. Love the fairy lights in the trees and general vibe of the area.
>
> (Age 35)

However, well-lit did not necessarily mean brightly lit. Instead, feelings of safety were supported by careful and considered lighting with minimal dark patches for hiding locations.

The analysis of lighting was predicated on this qualitative data. The value and deviation from targeted pedestrian reassurance studies in the past lie in the unprompted nature of how light appears as a crucial factor to people's experiences of safety at night in the *Free to Be* study.

Quantitative data set

The second part of the collaboration sought to create an understanding of how light is described and experienced by the general public through technical lighting terms to inform future design decisions.

The *Free to Be* data demonstrated that while light is a universally experienced factor of the night-time, the complexity behind how light can be designed and technically specified remains within specialist lighting knowledge. Moreover, light is an issue permeating compliance-based design that does not centre human experience and perceptions of light, instead focusing solely on the practicalities of vision, which too often results in poor night time atmospheric outcomes. This is reflected in the design of Australian night-time environments, where there is a tendency in designing for public spaces to choose a worst-case scenario by stakeholders to de-risk, and blanket the issue with more light (AS/NZS 1158.3.1:2020).

The overarching motive of the quantitative analysis of lighting was due to the frequency of lighting (38%) mentioned in 'unhappy' pins within the design of the physical environment (as opposed to the layers of the qualitative feedback that dealt with social matters). For example, these insights from *Free to Be* participants suggest the importance of light to the experience of the night-time in an Australian context for women and girls:

> Never felt safe walking in this area, even if I am not alone. The lighting is terrible and the design of the walkways leaves a lot of spots hidden from view.
>
> (Age 19)

> This area feels so dodgy. The light level means you can't see who is approaching you.
>
> (Age 39)

In order to further a more nuanced understanding of how lighting can affect perceptions of safety beyond bright/dark descriptions of light, the quantitative analysis was designed to measure a number of parameters beyond that of illuminance, the most common measure of urban lighting, the amount of light omitting from a luminaire source, measured in lux levels.

Field studies were completed across 84 sites that had been identified in the *Free to Be* study in Melbourne (Figure 5.1). Each site investigation worked with a framework of measurements based on a typical urban lighting design specification with LED luminaires and measured based on the P categories as described in AS/NZS

1158.3.1:2020. These measurements were undertaken by Christopher Alexander and Hoa Yang from Arup Specialist Lighting across consecutive week nights between 9:30pm and 4:00am in December, summer time in Melbourne.

Field measurements were conducted with typical measurement devices that lighting designers use to assess the existing or post-occupancy of the lit environment. This included an illuminance metre for illuminance and a spectrometer for colour rendering and

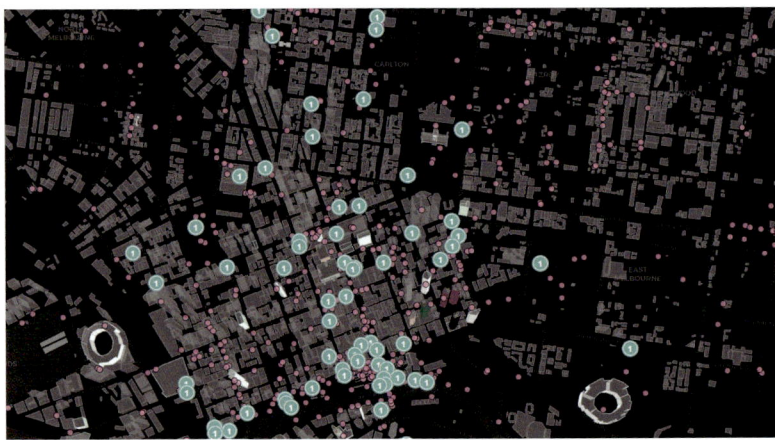

Figure 5.1
Screen capture of
the collected sites in
Melbourne in blue.
*Image: Hoa Yang, Arup
(2018).*

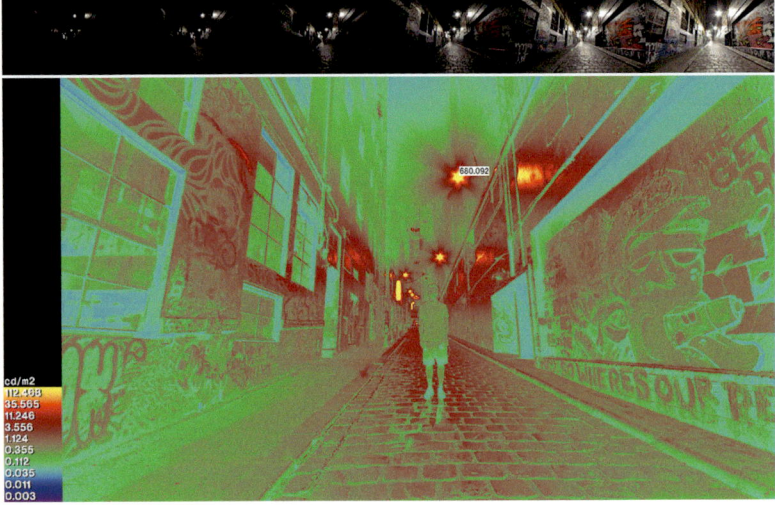

Figure 5.2
Stepped High Dynamic
Range images (top)
and false colour
output from radiance
(bottom).
*Image: Hoa Yang, Arup
(2018).*

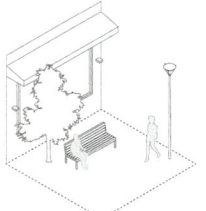
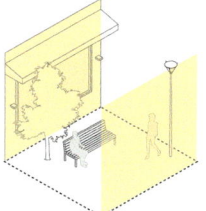
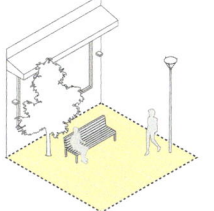
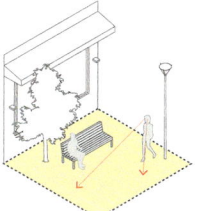

**Site
Typology**

Of the sites that were studied, there are proportionally significantly more safe pins than unsafe in the Civic Category.

There are proportionally more unsafe pins than safe pins in the Public Transport and Non-Arterial categories.

**Vertical Plane
Luminance Ratio**

Safer spaces had an average ratio between the left and right planes that is 10 times less than that of unsafe perceptions of space.

**Ground
Plane**

While there was not much difference between safe and unsafe perceptions in wide footpath widths, the findings show that small footpath width correlates to an unsafe perception of space.

**Escape
Route**

The presence of an escape route from the immediate path of travel contributes to a safe perception of space.

Figure 5.3 Key human factors findings.
Image: Hoa Yang, Arup (2018).

colour temperature. A further measurement of the perception of brightness (luminance) was measured to understand how light emitting from luminaires was interacting with the architectural context of each site. This was completed with High Dynamic Range (HDR) photography, a luminance metre, and post-processed in Radiance, a Unix-based backwards ray-tracing program that is an industry-leading software to simulate lit environments (Figures 5.2a and 5.2b). All measurements and analyses were based on the assumption of a person with good visual acuity.

To further understand the architectural and urban context of light and design, a human factors analysis was carried out on 15 'sad' and 15 'happy' sites. This broke down each site into a structured survey of 38 elements that aimed to address some of the social, cognitive, emotional, physical and cultural aspects of the site, determined by the technical Arup Human Factors engineer Chris Simmons. Elements in the survey included an assessment on the width of a footpath, the presence of escape routes and the presence of any foliage obscuring the pedestrians' view. This provided the quantitative analysis of light with some context and allowed for the exploration of an in-depth understanding of the convergence of light with the local spatial characteristics, use and material attributes, which can all affect one's interpretation of a space.

Findings and lighting practice

The initial findings from the data analysis revealed compelling correlations between current assumptions of light levels and unsafe spaces. It demonstrated that designing only towards the lighting standards misses the nuances of the experiential impact that lighting can have on perceptions of safety in spaces, challenging best-practice urban lighting design approaches in the Australian context.

The following technical aspects of light indicate a framework of key technical considerations when designing for perceptions of safety at night for marginalised urban experiences. The analysis of these parameters offers a practical approach and design framework to advance understandings of the connection between lighting, the urban environment and the perception of night-time safety for marginalised groups. The framework undertaken demonstrates how lived experience can be centered within a conventional and technical design language to inform practitioners.

Illuminance
Historically, research into the perception of safety and illuminance levels have found that high levels of illuminance make people feel safer (Peña-García et al. 2015; Rea et al. 2015; Blöbaum and Hunecke 2005). The analysis from the quantitative measurements in this study found that both vertical and horizontal illuminance correlate a higher

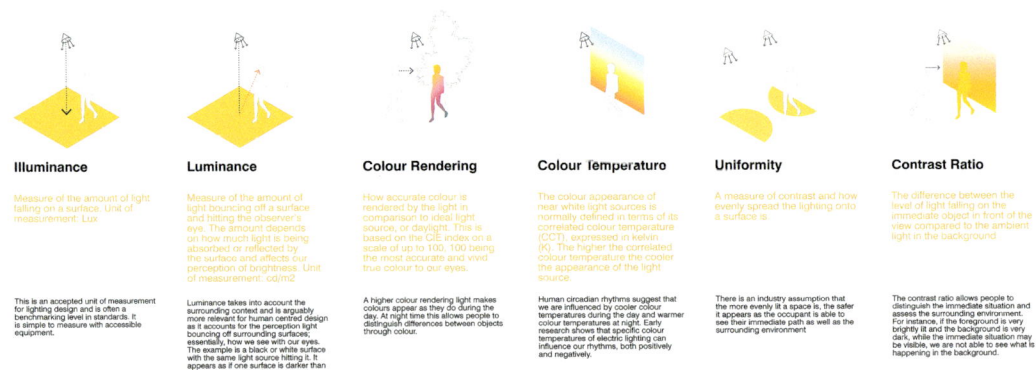

Figure 5.4 Key technical lighting measurements.
Image: Hoa Yang, Arup (2018).

level of light with unsafe perceptions of space. These findings support more recent research which suggests a correlation between lower levels of illuminance and perceptions of safety (Fotios et al. 2019). These findings are in contrast to the Australian urban lighting standards, which outline applying higher levels of light to improve perceptions of safety (AS/NZS 1158.3.1:2020).

Luminance

Research has demonstrated the importance of the perception of brightness as perceived by the human eye (luminance) – which accounts for the amount of light being absorbed by surfaces before bouncing back into the visual arena to create one's sense of brightness in an urban environment (Johansson et al. 2010). The findings from the human factors analysis alongside the analytics of luminance data in this study has indicated the importance of a balanced and considered approach to designing for perceptions of brightness, where the interplay of the light emitted from the light source with its surrounding surfaces are crucial considerations in evoking perceptions of safety for young women and girls. There is currently limited use of the measure of luminance in lighting standards, which focus heavily on illuminance, highlighting the often difficult exercise in capturing, extrapolating and post-processing captured luminance datasets.

Colour rendering

There was a strong correlation of poor levels of colour rendering in a wide range of spaces that are perceived as unsafe, while safe perceptions of spaces correlated more generally with a higher level of colour rendering. Current research on the impact of colour rendering for perceptions of safety has found that there is no correlation (Boyce et al. 2000), suggesting that further research is needed in this area. Typically applied only in art gallery and museum settings where colour accuracy is highly important, this study has found that colour rendering in an urban setting can acutely affect the accuracy of differentiating shapes and colours at night. The ability to provide accurate visual assessments of a night time scene has the power to shape behaviours as a result of perceived danger from passers-by and hidden objects. With the economic viability of specifying urban luminaires with high colour rendering being more accessible due to the development of LED and smart cities technology, the findings from this study suggest colour rendering as an important design decision to consider when vision is being augmented by electric light to improve perceptions of safety.

Colour temperature

A much smaller range of colour temperatures was found for sites with a safer perception of space, which were on average, spaces with a warmer colour temperature. This deviates from current lighting research comparing high-pressure sodium lamps (warm colour temperature) with ceramic metal halide (cold colour temperature), which concluded that 'people perceive areas illuminated with white light to be brighter, safer and more comfortable than [. . .] yellowish light' (Knight 2010) suggesting that colour temperature is also another area for further research. The findings from this study show that the application of warm colour temperatures can be likened to the inviting glow of a flame in the hearth of a home, to encourage long term dwell spaces that improve the comfortability and perception of safety in the city. In contrast, newly adopted LED street lighting technology has predominantly created cool colour temperature lit atmospheres in the Australian urban context, which evoke the sense that one should not linger, fostering transient spaces that people move through quickly.

Uniformity

Traditionally, research has found that a higher uniformity correlates with a perception of safety (Fotios et al. 2019). As such, standards often require levels of uniformity that is incongruent to the illuminance requirement (AS/NZS 1158.3.1:2020), resulting in an over-installation of luminaires to meet uniformity levels. The findings in this study show that there no noticeable difference in the uniformity levels between safe and unsafe perceptions of spaces. This suggests that uniformity, or how evenly the path is lit, may be less important to perceptions of safety for women and girls in Australia.

Contrast ratio

The findings for contrast ratio show that an imbalance between the foreground and background brightness is an indication that a space can appear unsafe. This may relate to dark areas in the background where people have ample space to hide and where there is a lack of brightness beyond the immediate path of travel so that it is difficult to assess alternative routes. While research has found that being able to recognise an incoming person's face as being a significant factor in the perception of safety after dark (Fotios et al. 2019; van Rijswijk and Haans 2017), the findings from this study suggest further research into the importance of seeing the entirety of an approaching person's body and overall scene, rather than just the ability to see the face.

Discussion of findings

Three key aspects of lighting in relation to perceptions of safety were found to correlate to a safe perception of space, offering opportunities for further investigation:

1. *Brighter is not always better.* Once light levels reach a certain threshold, higher levels of light appear to be perceived as less safe. The contradiction of levels of brightness required for perceptions of safety between existing bodies of urban night-time knowledge and the Australian lighting standards (AS/NZS 1158.3.1:2020) suggest the need for further investigation into the differences in gendered experiences in scientific studies of light. The

Figure 5.5
High Dynamic Range image from the site of an unhappy pin, Melbourne, Australia. *Image: Hoa Yang, Arup (2018).*

methodology built into this project provides a starting point to bridge qualitative and quantitative sets of data, historically considered separately, to inform night-time economies that might consider all possible users beyond dominant demographic groups in research or industry.

2. *Consistent and layered lighting is important to create an atmosphere of safety.* Physical features, such as trees, the colour of surfaces and the types of lighting in urban spaces, affect the way light enters the eyes and can influence the perception of brightness and safety in a space. The findings demonstrate that lighting designs where there are multiple light sources carefully considered within the architectural context, including surfaces with different reflectivity and level of mirror effect (specularity), should be approached holistically in the lighting design process to create safer urban experiences at night. This type of lighting application reduces the 'floodlit effect', the sharp drop-off of light beyond a person's immediate path and the potential for glare and contrast to blind and disorientate. The site-specific nature of the crowdsourced data of lived experiences has unearthed the need for designs to consider a convergence between the implicit effects of the urban environment, such as passive surveillance, with the explicit effects, such as the presence of light to create atmospheres of safety. A holistic consideration of the urban environment with regard to the perception of safety embeds design practices within wider contexts of the spatial, temporal and atmospheric interactions that can affect a person's experience of darkness in cities. Such expansive consideration within policy, practice and compliance paves the way towards shifting responsibility away from women and minorities needing to adjust their behaviours and movements to improve perceptions of safety at night.

3. *Quality over quantity.* Warmer colour temperatures and high colour rendering were found to be preferred in happy spaces. In the transition between sodium and metal halide luminaires into LED technology, many lit environments in Australia are still currently lit with warm colour

temperatures, but with low colour rendering. While the initial colour temperature creates a warm and welcoming atmosphere, this is let down by low colour rendering that prevents the ability to distinguish shapes and colour. This characteristic of lighting makes it difficult to distinguish a bush from a person walking towards one in the distance. With new LED and smart cities technology, colour rendering and colour temperature are two components that can easily be specified and allowed for in design, while legacy luminaires have less flexibility due to cost and maintenance. The push towards smart lighting with LED in cities, high qualities in colour rendering, controlled beam angles and colour temperature is now more accessible and customisable by designers and should be considered in the design of future experiences in cities after dark (Cities Alive 2015: 14). The ability to customize night-time experiences creates more urgency to centre participatory stakeholder engagement of marginalised lived experiences at the concept stage for all urban lighting projects. Listening to and understanding the experiences and the reasons behind what curtails people's engagement with the urban night-time may start to shift how the urban environment can be designed to facilitate an equitable night-time future that is accessible for all.

It is important here to recognise that environmental solutions to women's perception of safety after dark are only part of the equation. As Koskela and Pain (2000) have found, social and political relations that structure physical spaces, particularly the reproduction of patriarchal power relations in public space, have much impact on women's feelings of unsafety. That is, women's feeling of unsafety and avoidance of particular spaces at night are due to a 'fear of unknown men' or a certain reputation that places hold for being 'bad' or 'rough' (Koskela and Pain 2000: 275). As this study suggests, lighting is an element that has the potential to impact both the social and physical characteristics of space. As Painter (1991) contends, lighting can make a space feel as though it is well ordered and controlled, contributing to natural surveillance and making potential crimes visible to others. Consideration of lighting is just one component in the complex range of strategies required to tackle individual, environmental and social factors that curtail women's engagement with public space after dark. However, as a component of public space that is frequently pointed to as needing improvement in safety studies (Koskela and Pain 2000; Painter 1991; Shaw et al. 2013), localised approaches to lighting design have the potential to bring about positive environmental changes. Findings of this study reinforce that further research into the quality of lighting design can be an important mechanism to improve perceptions of safety during the night-time. This is coupled with the principal that women and other minority's expert lived experiences of the urban environment provide important insight for designers, planners and policymakers as to how these spaces might be improved.

Conclusion

While additional research is required to drive better insight into the role that lighting plays in the perception of safety in cities after dark, this research demonstrates that urban lighting standards and current practices are insufficient. The collaboration highlights the importance of incorporating voices outside of the dominant group of society and including usually ostracised experiences into the design process. To design truly inclusive and safe cities, design practice needs to centre the experiences of the outliers in society first. The framework presented in this study demonstrates a way forward in terms of how lived experiences can be integrated into design practice and compliance requirements for night-time spatial equity. It demonstrates that a designer-led approach of engaging with academic research can result in more equitable design outcomes that lie beyond merely meeting compliance requirements. This study is the beginning of a dialogue between designers and public, academic and policy stakeholders to recognise the importance that designing for the majority or most dominant voice does not result in inclusive environments. To gain a more equitable and inclusive insight into experiences of the city, co-design workshops to validate assumptions and data sets of other minority groups within the society should also be included in future studies.

References

Andrew, C. "Resisting Boundaries? Using Safety Audits for Women," in Miranne, K and Young, H (eds.), *Gendering the City; Women, Boundaries, and Visions of Urban Life*. Boston: Rowman & Littlefield (2000): 157–168.

Aruldoss, V, and Nolas, S. "Tracing Indian Girls' Embodied Orientations Towards Public Life." *Gender, Place & Culture* 26, 11 (2019): 1588–1608.

Arup. *Cities Alive: Rethinking the Shades of Night*. London: Arup (2015).

Beebeejun, Y. "Making Safer Places: Gender and the Right to the City." *Security Journal* 22, 3 (2009): 219–229.

Blöbaum, Anke, and Marcel Hunecke. "Perceived Danger in Urban Public Space: The Impacts of Physical Features and Personal Factors." *Environment and Behavior* 37, 4 (2005): 465–486.

Boyce, PR, Eklund, NH, Hamilton, BJ, and Bruno, LD. "Perceptions of Safety at Night in Different Lighting Conditions." *International Journal of Lighting Research and Technology* 32, 2 (2000): 79–91.

Brown, A, and Kristiansen, A. *Urban Policies and the Right to the City*. Nairobi and Paris: UN-HABITAT and UNESCO (2009).

City of Darebin. *Gender Equity and Preventing Violence against Women Action Plan 2019–2023*. Preston: City of Darebin (2019).

City of Melbourne. *Economic Impacts of Covid-19 on the City of Melbourne*. Melbourne: City of Melbourne (2020).

Committee for Sydney. *Safety After Dark: Creating a City for Women Living and Working in Sydney*. Sydney: Committee of Sydney (2019).

Community Council for Australia. *The Australia We Want. Second Report*. [online]. Canberra: Community Council for Australia. Accessed October 26, 2019. https://www.communitycouncil.com.au/sites/default/files/Australia-we-want-Second-Report_ONLINE.pdf

Fotios, S, LiAchenko Monteiro A, and Uttley J. "Evaluation of Pedestrian Reassurance Gained by Higher Illuminances in Residential Streets Using the Day – Dark Approach." *Lighting Research & Technology* 51, 4 (June 2019): 557–575.

Fotios, S, and Unwin, J. "Relative Weighting of Lighting Alongside Other Environment Features in Affecting Pedestrian Reassurance." In *Proceedings of the CIE Centenary Conference "Towards a New Century of Light"* (pp. 23–31). Sheffield (2013).

Fotios, S, Unwin, J, and Farrall, S. "Road Lighting and Pedestrian Reassurance after Dark: A Review." *Lighting Research & Technology* 47, 4 (June 2015): 449–469.

Freedman, EL. *History of the National Council 1946–1976*. Australia: The Illuminating Engineering Society of Australia and New Zealand (1976).

Fyhri, A, Hof, T, Simonova, Z, and De Jong, M. "The Influence of Perceived Safety and Security on Walking. Walk; Getting Communities Back on Their Feet." Cost Den Haag, Netherlands (2010).

Goulds, S, Tanner, S, Kalms, N, and Matthewson, G. *Unsafe in the City: The Everyday Experiences of Girls and Young Women*. Surrey, UK: Plan International (2018).

Grover, A. "Gender Perception of Safety in Urban Public Spaces: The Case of New Delhi." *International Journal of Arts & Sciences* 9, 4 (2017): 325–334.

Johansson, M., Rosen, M, and Küller, R. "Individual Factors Influencing the Assessment of the Outdoor Lighting of an Urban Footpath." *Lighting Research & Technology* 43, 1 (2010): 31–43.

Knight, C. "Field Surveys of the Effect of Lamp Spectrum on the Perception of Safety and Comfort at Night." *Lighting Research and Technology* 42, 3 (2010): 313–329.

Koskela, H, and Pain, R. "Revisiting Fear and Place: Women's Fear of Attack in the Built Environment." *Geoforum*, 31, 2 (2000): 269–280.

Mehta, A, and Bondi, L. "Embodied Discourse: On Gender and Fear of Violence." *Gender, Place and Culture* 6, 1 (1999): 67–84.

Menih, H. "Come Night-Time it's a War Zone: Women's Experiences of Homelessness, Risk and Public Space." *British Journal of Criminology* 60 (2020): 1136–1154.

Narendran, N, Freyssinier, JP, and Zhu, Y. "Energy and User Acceptability Benefits of Improved Illuminance Uniformity in Parking Lot Illumination." *Lighting Research & Technology* 48, 7 (November 2016): 789–809.

Nasar, JL, and Fisher, B. "Design for Vulnerability – Cues and Reactions to Fear of Crime." *Sociology and Social Research* 76 (1992): 48–58.

Nasar, JL, and Jones, KM. "Landscapes of Fear and Stress." *Environment and Behavior* 29 (1997): 291–323.

Pain, R. "Space, Sexual Violence and Social Control: Integrating Geographical and Feminist Analyses of Women's Fear of Crime." *Progress in Human Geography* 15, 4 (1991): 415–431.

Pain, R. "Whither Women's Fear? Perceptions of Sexual Violence in Public and Private Space." *International Review of Victimology* 4, 4 (1997): 297–312.

Pain, R. "Gender, Race, Age and Fear in the City." *Urban Studies* 38, 5/6 (2001): 899–913.

Painter, K. *An Evaluation of Public Lighting as a Crime Prevention Strategy with Special Focus on Women and Elderly People*. Hatfield, UK: Middlesex Polytechnic (1991).

Painter, K. "Street Lighting, Crime and Fear of Crime: A Summary of Research", in Bennett TH (ed.), *Preventing Crime and Disorder: Targeting Strategies and Responsibilities, 22nd Cropwood Round Table Conference*. Cambridge, UK: University of Cambridge (1996).

Peña-García, A, Hurtado, A, and Aguilar-Luzón, M. "Impact of Public Lighting on Pedestrians' Perception of Safety and Well-Being." *Safety Science* 78 (2015): 142–148.

Rea, M S, Bullough, J D, and Brons, J A. "Spectral Considerations for Outdoor Lighting: Designing for Perceived Scene Brightness." *Lighting Research & Technology* 47, 8 (2015): 909–919.

Roberts, M. "Engendering Urban Design: An Unfinished Story", in Staub, A (ed.), *The Routledge Companion to Modern Space and Gender*. Oxon and New York: Routledge (2018): 119–130.

Shaw, M, Andrew, C, Whitzman, C, Klodawsky, F, Viswanath, K, and Legacy, C. "Introduction: Challenges, Opportunities and Tools", in Whitzman, C, et al. (eds.), *Building Inclusive Cities: Women's Safety and the Right to the City*. Oxon and New York: Routledge (2013): 1–16.

Standards Australia. "Lighting for Roads and Public Spaces Pedestrian Area (Category P) Lighting – Performance and Design Requirements". *AS/NZS 1158.3.1:2020.*

Starkweather, S. "Gender, Perceptions of Safety and Strategic Responses Among Ohio University Students." *Gender, Place and Culture* 14, 3 (2007): 355–370.

United Nations. *Revision of the World Urbanization Prospects. Population Division of the United Nations Department of Economic and Social Affairs.* New York: UN DESA (2018).

Valentine, G. "Women's Fear and the Design of Public Space." *Built Environment* 16, 4 (1990): 288–303.

Van Rijswijk, L, and Haans, A. "Illuminating for Safety: Investigating the Role of Lighting Appraisals on the Perception of Safety in the Urban Environment." *Environment and Behavior* 50, 8 (2017): 889–912.

Vera-Gray, F, and Kelly, L. "Contested Gendered Space: Public sexual Harassment and Women's Safety Work," *International Journal of Comparative and Applied Criminal Justice* 44, 4 (2020): 265–275.

Wattis L, Green E, and Radford J. "Women Students' Perceptions of Crime and Safety in Negotiating Fear and Risk in an English Post-Industrial Landscape." *Gender, Place & Culture* 18, 6 (2011): 749–767.

Williams, R. "Night Spaces: Darkness, Deterritorialization, and Social Control." *Space and Culture*, 11, 4 (2008): 514–532.

World Bank Group. *Handbook for Gender Inclusive Urban Planning and Design.* Accessed April 5, 2020. https://worldbank.org/rn/topic/urbandevelpment/publication/handbook-forpgender-inclusive-urban-planning-and-design

Chapter 6

HOW THE CITY FEELS

Workshopping lighting design in public space

Shanti Sumartojo

DOI: 10.4324/9781003182610-6

Shanti Sumartojo

Introduction

How city spaces *feel* is a crucial aspect of urban quality of life. The sense of our surroundings as safe, threatening, lively, fun, dangerous, inclusive or scary shapes not only how we feel as individuals but also how our bodies, feelings and imaginations are situated in and through urban space. The implications of any of these feelings directly link to questions of equity, accessibility, fairness and justice, as well as the social and economic success (and even how success is defined) of our cities. The design of our cities plays an important role in how we make sense of and experience them, and as an aspect of urban design, lighting is a crucial element.

Lighting, however, because of its complex sensory, cultural and discursive affordances plays a special and distinctive role in how particular feelings configure in and through the built environment (Edensor 2017). Safety, anxiety, comfort, fear, frustration and more all tinge and accompany how we move in the city, as we slow or quicken our pace, lower or amplify our voices, and seek to take up space in particular ways that help us feel most comfortable. The role that light and lighting plays in these daily and ongoing micro-adjustments demands closer attention. Such an important design process often occurs in urban environments where lighting has been installed without fully appreciating the perspectives of people who will encounter it (Zielinska-Dabkowska and Xavia 2018). Despite the role of light in contributing to how spaces feel, the complexities of how these feelings are made, the different things that configure to give rise to them, the role of movement and stillness, and the varying perspectives that we all bring to urban space as individuals are difficult to account for, let alone incorporate into design.

This chapter seeks to address the complexity of experiential perspectives. It shows how design ethnographic research approaches to illumination can offer new ways to understand the experience of light and lighting in urban spaces, and the implications for lighting design processes and outcomes. I am particularly interested in how lighting design might contribute to and help to generate a sense of safety or vulnerability, and contend that such an approach is valuable because it can directly address how light and lighting help constitute how places *feel*. This is an urgent and ongoing question for designers, industry and governments alike as they design for and regulate our shared urban environments.

To demonstrate this way of researching and understanding lighting design, the chapter describes a co-design workshop that saw our research participants investigate the affordances of light in an inner-city laneway by manipulating a range of luminaires and reflecting on how their decisions shaped the feel of the space. The workshop approach we undertook was developed by the author with a range of collaborators, including lighting designers Hoa Yang and Tim Hunt from Arup, and Nicole Kalms, Jess Berry and other members of the Monash University XYX Lab, and built on previous collaborative investigations of lighting design that attended to the atmospheric qualities of urban space (Sumartojo and Pink 2018a, 2018b; Sumartojo et al. 2019). At the heart of this was the recognition that how we feel in the urban built environment is shaped by a continually emerging range of factors that lighting design is a part of and contributes to but cannot comprehensively determine. The chapter concludes by addressing the implications of this approach for both researching and designing urban lighting.

Light and lighting: knowing as you go

In this section, I provide two examples from my own collaborative research on light that have used techniques that attend to its sensory, mobile and affective qualities, aspects that I will return to in the discussion of the co-design workshop later. While these are accounts of particular methodologies, they also show how new concepts become available when research attends specifically to the relationality among built environments' intangible qualities, such as light, and people's individual and embodied feelings that these elements configure into.

In focusing on how people *experience* light, therefore, these examples are highly relevant for designers, even though they did not use the same terms or approaches that either UX designers or lighting engineers usually deploy. This research underpins the Melbourne workshop described later in this chapter because it shows, through the customisation of methods over several projects, the collaborative development of a distinctively design ethnographic approach to understanding how people make sense of their lit worlds, including lighting design (Pink et al. *forthcoming*). Both of the examples are forms of what we called a 'light walk' – a mobile, embedded and site-specific approach in which researchers and research participants moved through the city, attending to, photographing, videoing and discussing light and lighting. While discussion might happen during or after the walk, the research findings emerged as a direct result of affectively and visually sensing light and lighting, and reflecting on it either in situ or with visual materials that researchers or participants themselves made. The attention to movement that the light walk afford also emerged as important in the lighting workshop that is the focus of this chapter, as I will show.

The first example is from a 2017 study, when Sarah Pink and I asked participants to take photographs of automated light on one normal daily commute, in a study of everyday automation. We did not stipulate what these photographs should look like, what technologies they should include or where exactly they should be taken, instead inviting participants to decide this for themselves. We then video interviewed the participants with these photographs, asking them to explain what they photographed, why, and how they felt when they took it (see Pink and Sumartojo 2018; Sumartojo and Pink 2018b). This open approach meant that participants could nominate their own terms for how they perceived and understood automation and light. Even when something was not automated or when participants were not sure, they were able to use their images to explain how they decided this, which helped us understand how automation was understood in everyday life in general and how it was both reflected in and understood experientially through what we came to call the 'lit world'.

This approach revealed how participants never understood light as single-source but instead as a mixture of different sources, colours, reflections, movements and intensities, as we explained:

> illumination of various forms mixed and mingled with many other elements . . . creating feelings and impressions that were well beyond what the designer of any one of the light sources may have intended. Here the relationship between light and materiality was paramount – light often exceeds material boundaries, seeping into neighbouring spaces, bouncing off surfaces and angles in unexpected ways and mixing with other lights, sounds, smells and hapticities.
>
> (Pink and Sumartojo 2018)

These combinations composed a distinctive 'lit world' and connoted different feelings that emerged together with light in the daily context of everyday movements and routines. Using photo-elicitation in this way allowed for an extended discussion about the details of the spatial and emotional encounters with light. Videoing the interviews with research participants also meant we were able to revisit the discussions repeatedly and attend to the gestures, pauses, inflections and other forms of communication that go beyond words alone. This visual ethnographic approach (Pink 2021) was well-suited to a study of the lit world because of light's ocular-centrism but also because it helped us understand how light came together with feelings and understandings of their surroundings in people's everyday lives.

Our second example of a light walk saw three fellow researchers (me, Sarah Pink and Tim Edensor) use a body-mounted GoPro video camera, which I wore on a shoulder strap, to record movement and conversation as we walked through different lighting conditions in Melbourne's city centre. Our aim was to consider how light might help distinctive urban atmospheres emerge and, to do so, develop a dialogic methodology where the researchers were in conversation with each other and in response to the city spaces we encountered as we walked and talked. Our three-kilometre route took us from the edge of the city, with residential towers interspersed with large office buildings, along the river and its noisy bars and restaurants, through the central Flinders Street train station and into Federation Square, the main plaza at the heart of the city. As we walked, we discussed what lights we could see; what our impressions were of them; how they connected to what we knew about the history, plans or intentions for the city precincts; what feelings they engendered; and what sorts of activities they signalled. We noted and reflected on the transitions from one area to the next and the distinctive lighting design schemes that characterised these different areas (Sumartojo et al. 2019).

By making a video trace (Pink 2007) of this journey, we were subsequently able to listen to and reflect on our conversation and also see some of what we visually encountered. We were able to use this collaborative dialogic and auto-ethnographic material to consider how the atmosphere of the city at night constantly changed and reconfigured, and what some of the important elements were in giving rise to our different feelings about it. By analysing the trace of our movements and the accompanying conversation, light and lighting was revealed as absolutely crucial to engendering a sense of excitement, nervousness, annoyance, pleasure, safety and vulnerability. We were also able to reflect on how we had different impressions of the same places and why this was the case. Importantly, the feel of lighting always emerged in combination with other factors, such as the type and volume of the sounds we could hear; other people's activities happening around us; the things that light made it possible to observe; or the action of wind or river water adjacent to our path. Because of its role in signalling social, economic and cultural activities, light also opened a route to talk about a wide range of other things, such as the domination of the skyline by corporate advertising or the dull quietness of a city-edge residential neighbourhood.

Both of these examples show how 'light walks' can be used in different ways to research light, moving outside the lab and beyond the scale or survey. This approach investigates lighting design in the environments for which it has been designed but also in combination with the rest of the experiential world. It shows how the 'lit world', of which lighting design is a crucial, but not the only part, is actually a world that also draws in the multisensory, imagined and speculative modes of our spatial encounters, one where we do not always know what we will encounter or how we will feel about it. Finally, it puts the research participants at the centre of how light is defined and understood by asking them to consider their experience of light in their own terms, rather than in those that may be predetermined by the researcher.

Our light walks allowed us to access how precisely we 'know as we go', which is a particular affordance of both researching in movement and in the use of go-along video as a research methodology (Pink 2007). Video also makes research materials available for subsequent reflection and analysis, and enables the consideration of details that might usually slip by unnoticed in the normal routines of life. By attending purposefully to light in these studies, we were able to surface and explore the lit world and understand how important it is in how the city feels. In the next section, I discuss a public space co-design workshop where insights from these previous studies were able to be applied to a new, lighting design–focused research project.

Club Lane workshop, September 2019

A few dozen metres long, Club Lane is an access lane in the Melbourne city centre, sandwiched between an office building and private members club. There is garage access at the end of the laneway and a few doorways and large rubbish skips in it. It opens to a side street opposite a multi-storey car park, and skyscrapers rear up in the background. It is lit from a variety of sources, mounted on the surrounding buildings, that point down into the laneway, beams which mix with the bright ambient glow from the car park opposite and the nearby office and residential towers.

Our rationale for the workshop was to investigate how light was part of how people felt when actually in the laneway. To do this, we asked our research participants to arrange luminaires in ways that made them feel most and least safe. Rather than trying to arrive at a final design outcome, however, we were most interested in how people worked with the luminaires, how they manipulated and directed them, how light and feeling were co-constituted, and how people arrived at configurations that invoke particular feelings. We were also interested in how this might translate into lighting design and how people would approach the question of lighting for the particular feeling of safety when asked to do so.

The workshop itself was led by Tim Hunt and Hoa Yang from the Lighting Design Team at Arup in Melbourne, as was informed by Hoa's work on lighting design (see Chapter 5, this volume) and a series of projects in Monash's XYX Lab on women and girls' safety (https:// www.monash.edu/mada/ research/xyx). I designed and conducted the ethnographic research as part of the workshop. About 25 people participated in the two-hour workshop, all of whom identified as women, and none of whom were lighting designers themselves. The workshop started with some exercises in lighting design that Tim and Hoa had developed, and then they asked people to work in groups to light the laneway space. Each group was assigned a lighting designer, and there were also two people from the lighting manufacturers on hand to help manage the luminaires, explain what they could do and answer any questions. Together, these lighting experts were there to help 'translate' the participants' discussion into light and to help participants overcome any technical hurdles.

To attune participants to light and get them thinking about lighting design, Tim Hunt led a quick exercise with handheld torches in an adjacent laneway (see Figure 6.1), a technique developed in the previous public events he had led. This introduced some basic vocabulary about lighting design and also helped people feel comfortable manipulating light for themselves, as they played around with brightness, movement, colour and direction. This exercise primed our participants to 'play' with lights and, in doing so, to start to think

Figure 6.1
People consider the placement of torches to familiarise themselves with lighting design vocabularies around brightness, clustering and movement of luminaires.
Photo: Shanti Sumartojo.

about ways to explain the different effects of light and how they felt, vocabularies that continued to develop throughout the workshop.

We then moved to the adjacent laneway, where participants broke into three groups, were given a range of luminaires and were asked to light areas of the space in two distinct ways: to feel most and least safe. I moved between the groups with a video camera, sometimes recording what others were saying and sometimes participating in the conversation by asking them to explain what they were doing or why. In the rest of this chapter, I discuss some of the insights that emerged from this workshop about how light was contributing to the feel of the laneway and the implications for lighting design from these findings and of using this type of design ethnographic approach.

Lighting as obscuring

Throughout the workshop, light was perceived not just with the eyes but with a whole body that also experienced feelings, such as safety, nervousness, ease or fear. Anthropologist Ingold (2011: 11) addresses this in his discussion of how the body is the means by which we come to know our habitual surroundings:

> perception is the achievement not of a mind in a body, but of the whole organism as it moves about in its environment, and that what

it perceives are not things as such but what they afford for the pursuance of its current activity. It is in the very process of attending and responding to these 'affordances'. . . in the course of their engagements with them, that skilled practitioners . . . get to know them. Meaning . . . is drawn from these productive engagements.

Ingold's discussion of the perceiving 'whole organism' implicitly locates the body of a person in their surroundings where all the things around the body are understood in relation to what the person is doing or wants to do. This means that people come to know the world and the things in it through their actions and intentions, a process that Ingold calls 'habitation', and it is through this process that they become experts or 'skilled practitioners' in their surroundings. This discussion of skill helps frame the articulation of specific 'light vocabularies' discussed earlier, where participants brought their own experiences both of being in the nocturnal city and of manipulating luminaries in the workshop together with a set of lighting design terms that they learned in the workshop. The development of their own 'light vocabularies' continued, as I show next, as participants found ways to express their understandings of safety and light through a shared process of manipulating luminaires and discussing their impressions of the effects as they did so.

For example, a common topic for the workshop participants was that how light felt was strongly related to what it obscured and what it revealed. That is, light was not perceived or understood as somehow externally generative of, for example, feelings of safety, but rather that such impressions were contingent on a dynamic and constantly negotiated set of physical surroundings, of which light was a part. Thinking of light in terms of what it might hide shows how it is made sense of relationally and how it brings different things and feelings together in our surroundings. Because the capacity of light to obscure surroundings is important to feel safe while moving through them, this is akin to the 'productive engagements' with our surroundings that afford various activities that Ingold (2011: 11) discusses.

This requires us to move beyond thinking only of luminaires and their placement, brightness or colour, for example, and instead to think of them as part of a set of dynamic configurations of feeling, where light plays a role but is not necessarily determining. Instead, lighting's importance lies in how it connects the way we are able to sense our surroundings on the one hand and how we feel in and about them on the other. That is, what is revealed or obscured through illumination is the important thing, rather than the specific lighting schemes in themselves. This was expressed most clearly in discussions of shadows (see Figures 6.2 and 6.3).

In Figures 6.2 and 6.3, a group investigated a doorway corner by lighting it brightly and then stepping in and out of the shadowy recess to test where one group member became visible. As they did so, they altered the placement and direction of the luminaire, discovering through a series of small changes what could be hidden through the use of light, with the implication that even a shallow recess could hide someone if the shadow was dark enough. In this way, as a group,

Figures 6.2 and 6.3
Video screenshots of a research participant testing how they became visible in the shadow of a recessed roller door as they moved in and out of the light.
Photo: Shanti Sumartojo.

they made their own new understandings of how light felt through their manipulation of it, rather than knowing in advance what would feel safe in that particular space. It was through testing its capacities to reveal or obscure that feelings became known – put differently, it was their own investigations of light and lighting, and the reflection and conversation they shared with each other about it, that enabled the emergence of a new way to sense and describe feelings of safety or danger in the laneway setting.

Meanwhile, further down the laneway near a large rubbish skip (Figure 6.4), another group was also experimenting with eliminating shadows by moving light sources around. Although the physical surroundings were different here, the workshop participants similarly discussed and experimented with different levels of visual perception, discussing what they could see, what might be hidden and how these different arrangements felt. For this group, feelings of safety were related to the contrast between lighter and darker areas, as much they were linked to darkness itself; as Hvass et al. also discuss (Chapter 8, this volume), visual contrast between more- and less-illuminated spaces was an important factor. Here, being able to perceive with light did not simply mean more light, but rather it meant light that was sensitively designed in terms of visual perception in the specific built conditions of the site. One participant explained:

> the brighter the light, the harsher the light, the darker the shadows and that contrast feels a little bit, um, intimidating . . . and the darkness is very dark and that contrast of not being able to see where those shadows are, especially when the light is shining on the bins front on or side on, there's almost a halo of shadow around the bins which for me feel like the least safe area, that's where people would hide.

In this passage, we can see the continued refinement of a vocabulary of light and safety around contrast, brightness and shadow, but one that emerges directly from the things in the laneway with the research participant. She explains that while her sense of safety relies on what she can see, this is also contingent on how many things come together: the lights, the bin, the shadows that are cast and the bright reflections that create an obscuring contrast in the darkness of the alley. Light here does not stand on its own, but rather it is made

Figure 6.4
A video screenshot showing what one participant described as a 'halo of shadow' on the bin, potentially making it difficult to see anyone nearby. *Photo: Shanti Sumartojo.*

meaningful in its relationship to darkness and contrast, from which a feeling of safety is experienced or not. This is a complex and contingent account of lighting design that moves far beyond quantitative measurements that underpin many industry standards.

Lighting, movement and anticipation

How we imagine, anticipate and prepare for what might happen next are important and ongoing aspects of our experiential worlds. For example, in terms of how places might be felt as atmospheric, 'anticipation, foreknowledge and pre-existing views of different material and immaterial elements play a crucial role' (Sumartojo and Pink 2018a: 5). Anticipation is an process by which 'the imagination, because it anticipates and pre-views, serves action, draws before us the configuration of the realizable before it can be realized' (Sneath et al. 2009: 12). That is, our perception of the world is shaped by imaginative and anticipatory processes *before* things happen, rather than only *as* they happen. Light and darkness play an important role in this because they are part of how we make sense of our surroundings, not only in immediate terms but also in terms of what we imagine might be about to happen. To some degree, this anticipatory aspect underpins lighting design because designers and architects imagine, often with the aid of visualisation techniques, how light will look in order to plan the placement of luminaires or the detailed features of facades or materials, for example. However, the horizon of anticipation described by our research participants was not only much closer, it also constantly changed as they moved luminaires and themselves in the laneway workshop site.

Accordingly, the workshop explored different ways that light and anticipation come together. First, light is the means through which we can visibly apprehend and make judgments about our surroundings, part of the processual 'knowing as we go' of our environments. Second, anticipatory sense-making relates directly to how we *move* through the city, building on understandings of individual 'process[es] of thinking and knowing' that are 'formed along paths of movement' (Ingold 2010: S121). That is, not only are we seeing in light, but also 'our movements through the lit world are a vital part of how the city is experienced and understood, which both condition movement and condition our experience of it' (Sumartojo and Pink 2018b: 372). Light is often experienced as we move and as light sources appear to move with or past us. Moreover, light also signals constant movement around us, as advertisements of shop signs flicker or pulse; cars, trains or trams drive past; people move around us with illuminated screens; or we glide along streets ourselves in different vehicles, like trams, trains or cars. Usually with others, but even if alone, 'movements of all kinds are profoundly social activities that are both perceptive of the world and generative and transformative of it' (Vannini 2015: 3). In a moving world, anticipation comes together with light by way of decision-making about alterations to our paths when being able to see what is around us affords enough response time to change routes or slow or speed our pace to feel safer, or to move away from areas that feel less safe.

As in the earlier discussion of light's obscuring and revealing properties, there was implicit, ongoing work on the part of our research participants that enrolled light in anticipatory processes of feeling their way forward through their surroundings and assessing their personal safety as they went. During the context of the workshop, there was some experimentation with moving towards and away from luminaires or moving the light fittings themselves, as in the discussion of shadows and concerns about where people might be hiding. The response to a perceived threat to safety in terms of lighting design was to place luminaires so that the entire space

Figure 6.5
A video screenshot showing a research participant gesturing to indicate how she would prefer to see further ahead into the laneway space.
Photo: Shanti Sumartojo.

could be scanned at a glance to enable a quick assessment of its safety. Even if the ground surface was uneven, safety was felt as the capacity to see as much as possible of the whole laneway space (Figure 6.5) rather than just what was underfoot, as a research participant explained:

> Personally I'd much rather be able to see further in front than right at my feet because then I can see who's coming, what's coming, where the end [of the laneway] is, if I can only see my own feet then you've still got all this darkness around you, it's not very helpful.

Here, safety and lighting were bound together in how quickly she might be able to move away or towards places or people, how much time there was to respond and how she assessed her surroundings and made decisions on the move. The point here is that lighting has affordances that play out in movement and across time, and so lighting design needs to account for the quick scans or glances, the changing perspectives of the walker or rider in motion, and the different shadows, illuminations, visual blockages and lines of sight that constantly change around us as we move. Moreover, this movement is accompanied by ongoing and powerful imaginative processes that constantly anticipate what might be encountered, as a means to stay safe.

Lighting as care

As we have discussed, if safety is a feeling that is made in specific spatial configurations of which lighting is a part, then this affective impression is constantly made and remade in the particular and mobile conditions of the city at night. A related impression, and one that emerged as we reflected on and in the laneway setting, was that lighting is bound up in notions of *care* for and in the city – that is, lighting that feels both purposeful and safe was interpreted as expressing a welcome sense of care that decreased nervousness about dark urban spaces. In the context of urban settings, this might be interpreted through the impression that money or attention had been spent on a particular site, suggesting that it was valued, or other purposeful demonstrations of care that tell people that a site and the things that happen in it are important. Particular places might enable people to show care for one another or simply engender a feeling of being cared for because the built environment is understood to have been tended, curated, maintained or designed with attention to detail and empathy for how people will experience it.

Addressing this explicitly, Vaughan (2019: 8) advocates for design as a practice of care, recognising that 'there is now a recognition that it is real people, with all their complexities and differences, who are at the heart of our design acts'. She describes 'design as a practice of care': a relational practice between the designer and other people, where designers empathise with and imagine the people they are designing for. Here, care is not only intended by the designer but is also *experienced* through design that has 'meaning for those being cared for' and that is purposefully generated by a designer with both the skills and the capacity to 'perform the actions of care' (Vaughan 2919: 12). In other words, design can *be* a form of care, in contexts far beyond traditional 'care' settings, like hospitals or schools, if the impact on people who experience it is positive or is experienced as caring. Even in a setting like Club Lane, care can be enacted and experienced through lighting design that helps support feelings of safety; and indeed, safety and care are impressions that often arise together.

Imrie and Kullman (2017: 2) make a related point when they ask how the 'skills and sensibilities of caring can be expressed through design practice in order to enhance the conviviality and well-being among those who inhabit, and depend on, cities'. Exploring the question of urban form and the fit (or not) between people and the urban forms they inhabit, they highlight the 'problematic encounters between people's bodily capabilities and built form' as a question of ethics (Imrie and Kullman 2017: 7). Lighting design, and its relationship to urban form, is important here because it is located precisely in the 'fit' between people and the built environment. In other words, because it is a means by which people both make sense of and understand their urban surroundings, lighting is one way that care can be expressed in the city through distinctive effects that 'guide understandings about how to act and manoeuvre within distinctive spatial and social settings' (Sumartojo et al. 2019: 2).

For many of our workshop participants, there was a sense that if places are 'cared for', they are also more pleasant, which contributed to a sense of safety. The impression that places are cared for was related directly to forms of lighting design because of its capacity to highlight attractive architectural features or material textures; because it directed light in ways that avoided shadowy recesses or the obscuring glare of contrast; or because it simply looked 'expensive' and, therefore, communicated that it was valued by the owners or the council. Such understandings in our workshops were reached by arranging lights to feel most and least safe, with 'care' often mentioned as how a sense of safety was able to emerge in place, through lighting. In this way, lighting design can evidence care for a space, as related to feelings of safety, as one workshop participant explained:

> [If] it's lit with a sense of care, that someone's paid attention to the lighting scheme, to the lighting of the walls in a special way. It feels like someone's paying attention to the space, and that kind of conveys a sense of safety. People are aware of the space and it's not an afterthought. . .

Others identified the intentionality and aesthetic value of lighting design – beyond simply lighting for brightness – and told us that lighting design that appears intentional is reassuring, precisely because it signals care:

> Participant 1: When you can see that there's been some intent
> to do something interesting with light, it automatically

puts you at ease I think, because you know that someone's put some time and effort into . . .

Participant 2: . . . care for the space.

Participant 1: Yeah, exactly.

In this context, participants discussed how the descriptor 'designerly' meant that lights appeared to have been placed purposefully to high-light particular architectural features and included the use of pleasing repetition, colour and variable brightness. 'Designerly' schemes were perceived as more attractive and amenable, with the clear intentional-ity interpreted as a form of care for the site. Entangled in this was a sense that professional lighting design could be recognised by certain distinct visual features and that its practice was linked to feelings of care and, by extension, safety. The impression of something *having been designed,* in other words, connected to care through value and intentionality.

Care was also connected to a sense of responsibility. Pur-poseful lighting design signalled that someone was responsible for a site and what might happen in it. Lighting could, therefore, enable an ongoing sense of being cared for that seemed to move beyond the moment of design or of the person's encounter with design. This connects with Vaughan's (2019) proposition of design as a practice of care, in the sense of the ongoing qualities of both design practice and how it is experienced when it is articulated in the world. A care-ful practice of lighting design might, therefore, be able to move forwards with people, supporting feelings of safety in the moment of encoun-ter but also signalling intentionality, responsibility and attention that conveyed an *expectation* of safety into the future. Connecting to the discussion of anticipation earlier, this quality can also support feelings of safety because purposeful and care-ful lighting can decrease the need to constantly scan one's surroundings for potential threats.

One expression of this was that lighting design could com-municate the sense that the place would be maintained, was pos-sibly subject to surveillance or was a high-traffic area where there would probably be safety in numbers. This extended to the likelihood of people taking care of others and an impression that someone would come and help if it was necessary. In other words, this feel-ing of care – and by extension, safety – was experienced and imag-ined as anticipatory, as discussed earlier. This helped with the small assessments and decisions that our research participants made as they traversed the city, as one person explained:

> I think if you're making a decision about whether to go down a laneway at night, you're trying to assess whether . . . how likely it is that you'll be the only person in that laneway with potentially someone dangerous. And ideally someone else might walk through and that makes you feel a bit safer, like there's more people. So if it seems more cared for, it feels more likely that someone's likely to help, to come to your help.

In these examples, demonstrable care on the part of the designer or the people responsible for city spaces was interpreted through lighting design that felt purposeful, sensitive and beautiful. This might be very minor, such as a slight angling of a luminaire so that it did not shine directly into the eyes of people entering the laneway or the placement of a light source in an alcove to both illuminate the recess and highlight the attractive texture of the brick wall. In other words, lighting design did not need to be large-scale, expensive or dramatic to be interpreted as caring. It did, however, need to appear empathetic, sensitive and purposeful to help support feelings of safety.

Implications for lighting design

Our co-design workshop demonstrated in very practical terms how lighting contributes to how the city feels. As a methodology, it showed what can be learned through asking people to engage in practices of making together and the rich reflections, discussion and insights that can result. In this case, the thing that was being made was an arrangement of lights, but along the way, by exploring the mechanics of obscuring and revealing, and by engaging in a world in movement where people are constantly exercising their anticipatory capacities, we were able to surface understandings of lighting for safety by way of concepts of care and responsibility.

We asked our research participants to engage in a simple task that allowed them to take control of luminaires and explore them on their own terms and in relation to their own lived experiences. By moving the research site into an inner-city laneway environment, a place where safety is important and where personal risk is constantly assessed, we were also able to show how complex feelings, such as safety, are made with and in particular surroundings. That is, these feelings always emerge in relation to where we are and what we are experiencing and thinking. An implication is that lighting design, because it is so entangled with feelings of safety, needs participatory and open approaches to investigating people's experiences of the lit world and for evaluating the effects of lighting design itself.

An example of what can be learned through this method is how particular lighting vocabularies developed as participants practiced rough, experimental and vernacular forms of lighting design for themselves. They came to understand light and lighting design through the collaborative manipulation and arrangement of luminaires, a process that started with their own previous experiences of many different conditions of the lit world. As experts of their own experience, our research participants were able to build on this to imagine how it might be possible to improve the feel of the city through sensitive and intentional lighting design. This differs from the laboratory settings of many evaluations of lighting and shows what can be learned when a design ethnographic approach is employed, one that locates lighting design in the settings where it is encountered and where it's careful deployment might improve the urban experience.

The practice- and experience-based language of lighting I have discussed here, and the distinctive ways in which it composes urban spaces, offers a subtle and responsive account of the relationship between lighting design and public safety. It shows how light and lighting are made sense of *in situ*, adding complexity to predictions about what lighting design will look or feel like before the moment of encounter. An implication is that lighting design needs to be driven not only by practical concerns and regulatory stipulations, but by concepts that go beyond visual apprehension and that connect to ideas of care, safety and belonging. Lighting both signals and reinforces who the city is for by way of how the city *feels*. Helping it feel more inclusive and safer, as I have discussed in this chapter, is a matter of care for all of us.

Acknowledgements

Many thanks to all the research participants who shared their time and experiences, and to Tim Hunt and Hoa Yang from Arup and Michaela Sheahan from Hassell who helped make the workshop possible.

References

Edensor, T (2017) *From Light to Dark: Daylight, Illumination, and Gloom.* Minneapolis: University of Minnesota Press.

Edensor, T and Sumartojo, S (2015) Introduction: Designing Atmospheres. *Visual Communication* 14(2): 251–266.

Imrie, R and Kullman, K (2017) Designing with Care and Caring with Design. In *Care and Design: Bodies, Buildings, Cities.* London: Wiley, pp. 1–17.

Ingold, T (2010) Footprints through the Weather-World: Walking, Breathing, Knowing. *Journal of the Royal Anthropological Institute* 16: S121–S139.

Ingold, T (2011) *Being Alive: Essays on Movement, Knowledge and Description.* London: Routledge.

Pink, S (2007) Walking with Video. *Visual Studies* 22: 240–252.

Pink, S (2021) *Doing Visual Ethnography* (4th ed.). London: Sage.

Pink, S, Fors, V, Lanzeni, D, Duque, M, Sumartojo, S and Strengers, Y (forthcoming) *Design Ethnography.* London: Routledge.

Pink, S and Sumartojo, S (2018) The Lit World: Living with Everyday Urban Automation. *Social & Cultural Geography* 19(7): 833–852.

Sneath, D, Holbraad, M and Pedersen, MA (2009) Technologies of the Imagination: An Introduction. *Ethnos* 74(1): 5–30.

Sumartojo, S, Edensor, T and Pink, S (2019) Atmospheres in Urban Light. *Ambiance.* Online 20 December 2019. http://journals.openedition.org/ambiances/2586

Sumartojo, S and Pink, S (2018a) *Atmospheres and the Experiential World: Theories and Methods.* London: Routledge.

Sumartojo, S and Pink, S (2018b) Moving through the Lit World: The Emergent Experience of Urban Paths. *Space and Culture* 21(4): 358–374.

Chapter 7

AT THE MARGINS OF ATTENTION

Security lighting and luminous art interventions in Copenhagen

Mikkel Bille and Olivia Norma Jørgensen

DOI: 10.4324/9781003182610-7

Introduction

Urban lighting has mostly been based on safety, functionality and amenity to make spaces visible, in general leading to increasing light levels throughout the world. In recent decades, however, new lighting technologies have allowed urban spaces to be lit in radically new ways with improved lighting quality and higher energy efficiency. The new technologies have enabled lighting designers to test the limits of formally approved standards for old lighting technology to show new ways of illuminating for the users (cf. Ebbensgaard, 2020). With the simultaneous emergence of a more consolidated lighting design profession, more urban spaces are now intentionally designed in elaborate artistic ways to create spatial identities after dark. Sometimes the lighting design aims to be a spectacular centre of attention, while at other times, it aims at subtle intervention without standing out as such but still affecting the feel of the place. With this turn in lighting design, particularly in Scandinavia issues around safety are still present, of course, but often act more as a backdrop than the main priority. The common priority of lighting for visibility – generally marking the 20th-century lighting practice of making things visible, including on CCTV – is now just one of several ways of shaping a lit world.

In Denmark, these new possibilities for urban lighting design are particularly employed in areas with social challenges. Lighting design interventions in these areas have traditionally been dominated by design narratives of 'opening up' visibility by removing green shrubbery and adding security lighting making selected areas fully visible. These areas are now increasingly redesigned with the aid of artistic lighting. In technical terms, since the early 21st century, this has included spotlights with a variety of colours and patterns creating high contrasts on surfaces, and movement sensors and automated lighting that changes the lighting over the course of the dark hours, shaping more aesthetic scenes on urban life than the common, even spread of light (Figure 7.1). The new technologies have made it fully apparent that urban lighting design is not just about making spaces *visible*. It is about making them *felt*.

This chapter addresses lighting as part of urban atmospheres through an ethnographic case study of a square in Copenhagen called Blågårds Plads. The square has, in recent years, been at the centre of several lighting interventions as part of the area's renewal policies,

Figure 7.1
Urban lighting design
in Northwest Park,
Copenhagen.
*Photo: Ida Lerche
Klaaborg.*

mixing both new, artistic and more conventional approaches to light-ing in a range of different light fixtures. Firstly, two places on the square have two clear bright lights spreading from fixtures hanging on the wall as part of the security lighting (in Danish *tryghedslys*, which refers to lighting that shapes a sense of secureness) to curb crime and fear. Secondly, under a few lines of trees and benches surround-ing the central part of the square, two different lamp designs with a total of 17 lamp posts offer small spreads of bright light. Thirdly, there are four lamp posts around a central sunken pitch with up to six fix-tures on each, spreading a mix of spotlight and snowflake patterned light (Figure 7.2). Fourthly, there is an area around a public library that has recently served as an urban stage for artistic lighting interven-tions, which we turn to in the end. And finally, there is the 'vernacular' lighting shining from homes, offices, shops, cars and bikes, adding to the luminous life at the square. It is, thus, a square that has both areas with bright spreads of light and other areas embracing gloomier or more artistic aspects of the urban night (cf. Edensor, 2013).

The square's lighting strategy is closely orchestrated in rela-tion to a small group of young men either associated with the criminal gang Loyal to Familia (LTF) or just hanging out with them. They stand on the south-eastern corner of the square near a kiosk and pizzeria

Figure 7.2
Lamp posts with
spotlights and
snowflakes.
Photo: Mikkel Bille.

most days, either around a visiting car, on bikes or just standing chat-
ting to each other. Any design intervention on the square, including
lighting, must in one way or another take their presence into consideration. The urban design in general has elements of security by

Figure 7.3
Security light by the
entrance to the square
near Apotek bar.
Photo: Mikkel Bille.

design (Figure 7.3), while simultaneously also engages lighting designers, technicians and artists in lighting design installations (Figure 7.4)
to create more atmospheric light beyond a surveillance paradigm of
thinking (cf. Otter, 2008). As an effect of the gang-related presence,

Figure 7.4
Artistic lighting design
Universe by Karoline
H. Larsen.
Photo: Light Bureau.

the square is marked by a substantial police presence watching this group and, in terms of lighting, the presence of bright security lighting (*tryghedslys*).

However, another story of the square is one of bars, restaurants, shops and a library that shape a markedly different vibrancy only a few metres away from where the young men are located, making the square feel very different depending on what one is doing at the square and where. There is not *one* atmosphere but many that interlink and fluctuate, and lighting is part of this atmospheric diversity and vibrancy in the dark hours. It is this other version of the square (along with the surrounding neighbourhood) that reached international fame in 2020 on *Time Out*'s list of coolest neighbourhoods in the world for its 'never-sleeps sort of atmosphere'[1]. Here, amidst policing strategies curbing crime, women walk with strollers, dogs are exercised, students commute on bicycles, children play, youngsters play football, homeless people meet and the occasional cultural event draws people in. For this chapter, then, what is essentially at stake at the square is how different atmospheres unfold alongside each other and, at least in the dark hours, are influenced by urban lighting.

To understand the role of lighting at a square with such diverse uses and atmospheres, we need to engage with how people understand and position themselves within urban spaces and the social values embedded in living and using this neighbourhood. The chapter is based on an ethnographic study of the square running for six months from September 2018. During the fieldwork, we interviewed nine stakeholders who worked at the square (police, designers and social workers), three residents along with people visiting the square for leisure. The starting point was the observation that the square *feels* markedly different depending on where you are positioned. This does not mean that one can essentialise exactly *how* it feels in certain locations at all times, but it allows researchers, designers and users to recognise and explore how a space may feel different depending on where you are and when. From inside a home, it may be cosy to see the rain fall on windows and the square, reflecting the street lights. From an *in situ* position at the square, this may be quite distressing, while people in offices at ground level may get a different sensation of the square in the evening rain. Through semistructured interviews, participant observation and informal conversation, we obtained perspectives both from people going to and from the square, and from people living their everyday lives on and around it. We believe that such an ethnographic perspective is useful for (lighting) designers because it highlights that the social life of light is not only its effect on visibility but also its impact on atmospheric qualities of spaces as embodied sensations of being, seeing and moving in them. Such sensations are naturally guided by the individual biographies and physiques of people but also users' and residents' adherence to more collective values and norms. For lighting to play the role that its designers intended, we need to understand and engage with the diversity of atmospheres, both desired and undesired, that people, lighting, urban design and usage co-create.

After briefly addressing recent literature on lighting and atmospheres from social sciences, this chapter outlines how a multiplicity of uses and connotations of Blågårds Plads centre around a local sense of a 'fighting spirit' and the co-existence of diverse groups of residents and visitors. It then delves into two distinct topics in which lighting plays a role in shaping Blågårds Plads's distinct atmospheric qualities: firstly, the impact of security lighting on the experience of gang presence among residents and users; and secondly, how artistic lighting draws on local notions of co-existence rather than security to shape an urban design intervention.

Mikkel Bille and Olivia Norma Jørgensen

Security lighting and atmospheres

Historically, lighting has held a central role in making spaces visible and making public spaces accessible to policing (Otter, 2008; Schivelbusch, 1987). Cook and Whowell, for instance, note that it is clear that 'visibility and invisibility are central to the policing and social control of public space' (2011: 618). Relevant for the gangs in our case, they note that 'Not only is policing about making the bodies and technologies of policing selectively visible and invisible, it is also about making 'troubling' populations selectively invisible and visible in public space' (2011: 613).

These various sorts of visibility have been central to the practice of using light to deter crime. In a meta-study of 13 studies of the effects of improved street lighting on crime, Welsh and Farrington argue that while the relationship between light and crime prevention is not straightforward, their meta-review nonetheless indicates that 'improved street lighting significantly reduces crime' (2008: 2). Why this is the case, however, is up for debate, they note, as there appear to be two main theories to explain why the reduction in crime occurs. One rests on a 'surveillance' paradigm, where increased visibility and number of people on the street increase deterrence of crime. The other theory rests on a notion that the improved lighting 'signals community investment in the area and that the area is improving, leading to increased community pride, community cohesiveness, and informal social control' (2008: 2). From these two theories, Welsh and Farrington most notably found that *night-time crimes* did not decrease more than *daytime crimes,* leading to the suggestion that the latter 'social' theory may be more plausible than the former surveillance theory by increasing community pride and informal social control deterring crime *both* day and night. While Welsh and Farrington interrogate what *more* lighting does and suggest why, their aim is not to dwell on the quality or type of lighting.

Lighting technologies are evidently important in shaping feelings of safety in urban spaces during the dark hours, and this is no less so in spaces that are popular in a night-time economy. Surveillance or visibility may be the tool to increase safety, but it comes with different effects, as van Liempt and van Aalst note: 'For some, more surveillance can lead to feelings of certainty, security and trust because immediate action can be taken. For others, it has a negative effect on the atmosphere and increases feelings of mistrust and insecurity' (2012: 291). Or as Cook and Whowell (2011) also critically note, there may be a preoccupation with the visual signifiers in research on social control and policing of urban spaces, leaving aside the complex multisensuous ways in which urban encounters are evoked, adapted and negotiated. That is, rather than engaging with whether one can see or not, there are atmospheric aspects that go beyond the purely visual and involve how that visibility makes one feel and shape a social sense of places. This means that what matters more than simply casting (more) light onto a surface based on *functionality* is *how* the light is designed in more qualitative terms to be *felt*.

Increasingly, social science research is turning to such issues of the atmospheric qualities of lit space. Ebbensgaard (2019), for instance, has shown how there is a marked difference between *seeing* light and being *in* it (see also Casciani, 2020). Working with older citizens, he has shown how new LED lighting technologies have shaped a nocturnal urban landscape that people need to make sense of, including through bodily exposure to the unlit or more dimly lit areas. Taking a couple of older women to an alleyway they normally avoided due to low light levels, Ebbensgaard notes that, 'When standing *in*, and seeing *through* the light, it appeared brighter than when standing afar looking *at* the light' (2019: 100). While the clear view offered by the new LED lights made urban spaces look different and more visible – and according to other research *as such* make people feel safe (Peña-García et al., 2015) – there are also potentials for understanding how the specific positions of the light spread and light from other places than public sources, such as shops and homes. Such sources may add 'an indispensable and affirmative domestic atmosphere to the nocturnal city, which makes people see better, feel safer and more confident walking alone without feeling lonely' (Ebbensgaard, 2019: 103). Low light level, in essence, may, under the right conditions, foster a sense of domesticity in the urban night and, through this, a sense of safety. Again, as with Cook and Whowell (2011), experiencing the urban night is not simply a visual engagement of looking *at* something but a multisensuous encounter where preconceptions along with actual visual experience *in situ* affect the experience.

This affective embodied sense of being in place and feel the atmosphere has gained increasing attention in recent decades of social science (cf. Bille et al., 2015; Sumartojo & Pink, 2019;

Schroer & Schmitt, 2018). Atmospheres denote the felt qualities of space that may capture the perceiver but may also be somewhat difficult to define or articulate. They are in a phenomenological tradition seen as the co-presence of subject and object (Böhme, 1998: 114), as that 'in-between' that shapes the experience of the world not reducible to either individual psyche or external world. Or as Ben Anderson comments, 'they are impersonal in that they belong to collective situations and yet can be felt as intensely personal' (2009: 80). Atmospheres take shape, on the one hand, through the affective qualities of architecture, objects and people as well as less tangible phenomena, such as weather, light, sound, smell and so forth. Lighting is a particularly efficient atmospheric technology, as witnessed in cinematographic manipulation of light to reach an atmospheric effect (Böhme, 2013; Groening, 2014; Spadoni, 2020). Yet the atmosphere induced by urban design is also felt by an attuned person that may be absent-minded and insusceptible to such qualities or may be so charismatic that they are able in return to change the feel of a place. Architects, designers and cinematographers are, in other words, not sure that the atmosphere aimed for will seize the perceiver in the intended way, and this is even less so in an urban setting where many different actors, politics and affective interests compete (Allen, 2006; Stephens et al., 2017). They may change rapidly or linger as an almost integral part of the spirit of a place (Norberg-Schulz, 1980).

In essence, then, atmospheres are felt yet fluid and ephemeral (cf. Degen & Lewis, 2019). While this obviously entails quite a few difficulties for researchers and designers alike – such as the problems of 'defining', creating or 'capturing' them – they are nonetheless a central part of (urban) life. Sumartojo and Pink (2019: 11–12) have argued that one way of getting around the different facets of atmospheres is to think of atmospheres as something to know *in*, *about* and *through*. Knowing *in* means for researchers to seek to enter into participants' worlds, sensing their point of view. Knowing *about* revolves around the process of representing something non-representational that in many ways avoids precise articulation or about what *is* articulated but may be felt in different ways. Knowing *through* helps us as researchers with an entry point to understand experiences and themes that may not at first seem connected but which actually affect people's world view. Thus, when asking our interviewees questions about atmosphere, we get answers about light but also social values and conceptualisations of space beyond what is merely visible or not, or normatively 'good' or 'bad'. By addressing atmosphere, in other words, we find a way into a lived version of lighting.

A melting pot with fighting spirit

The rectangular Blågårds Plads square is located at Inner Nørrebro close to Copenhagen city centre. The neighbourhood surrounding the square is what can be termed a 'melting pot' in terms of people, architectural styles and urban planning. Not only the square but the entire neighbourhood is characterised by being a place to hang out for everyone from families with children to gang members and homeless people. The square and the rest of the neighbourhood have a long history of housing many different nationalities, social classes and religious convictions, and a place where vibrant street life mingles with the presence of drugs, addicts and gang violence, leading to several municipal-led area renewals (e.g., Grönlund, 2014). In recent decades, it has also been turning into a place where people from all over the city go to hang out and drink coffee and beer, illustrated by the international recognition by *Time Out*.

The central part of the square consists of a sunken pitch, with benches around the edge and large trees on either side. At the north-western end, there is a playground, surrounded by public housing mostly inhabited by non-ethnic Danes, and which, until 2013, has been on the so-called 'ghetto list' of public housing with a high percentage of unemployment, immigrants and prison convictions, among other criteria. On the ground floor is a housing office as well as two passages with new light art installations, *Tracks*, that lead out to other parts of Nørrebro (Figure 7.5).

Along the western side of the square is a library on several floors, with large panoramic windows facing the square and plants growing up the building wall. On the corner of the building is a small community office. Blågårdsgade in the east is a street always busy with pedestrians, cyclists and sometimes car traffic. On the ground floor, there is a kiosk, a kebab shop, a pizzeria and two cafes. On the other corner of Blågårdsgade is the Apotek pub, whose clientele consists of both regulars who sit drinking tea, coffee or beer, and 'tourists', as one of our interviewees called non-regulars visiting from

Figure 7.5
Tracks by Karoline H. Larsen.
Photo: Light Bureau.

other parts of Copenhagen. Opposite the library is a yarn shop, a baby clothes shop and an alley leading to a church, as well as an Asian restaurant and a potter. In essence, the square enjoys a highly diverse and multicultural mix of shops, cafes, people and cultural offerings that ensure a constant presence of people.

Like many other places in Copenhagen, the area has witnessed the influx of an economically well-off middle class in the last decades. Observing at the square, you may find at one end a group of middle-aged people drinking a beer on a bench next to homeless people and, at the other, small children playing together in the sandbox. This heterogeneity is characteristic of the square. But the differences in income, political observance and nationality have never pulled people apart as such, as all our interviewees who frequently use the square would point out. A central figure in the public housing, Christian, described it as a 'fighting spirit'. Demographic changes notwithstanding, the residents are largely characterised by this fighting spirit, he argues, leading to public narratives of Blågårds Plads as a place where people gather for social protest from all around the surrounding neighbourhood of Nørrebro. He noted that 'There is a lot of the 70s intellectual elitism in the fighting spirit, and many of the children of the anti-authoritarian left-wing, broadly speaking, also come here. They are here for protests and vandalism towards capitalist corporations'. The fighting spirit is characterised by gathering both residents from the housing association and members of anti-authoritarian and left-wing groups during moments of protest. The narrative of both being a left-wing area and an arena for riots, in turn, has also made it a place for the opposing political spectrum to meet and spread their viewpoint, often leading to clashes. The image of the area is, thus, characterised by what seems like an incompatible vibrant urban life mixed with riots and violence.

Although many residents may also stay away from the violent activities, they recognise them as part of the spirit of the place. The fighting spirit was actually conceptualised in the initial design of the square revealed in 1918 with an underlying intention of creating a square with room for popular meetings and popular assemblies. This design forms a backdrop to more recent efforts to curb crime and address feelings of lack of safety in the area. We were constantly hearing descriptions of this assembly and fighting spirit from our interviewees who also underscored a notion of co-existence: 'There is room for everyone', 'Blågårds Plads is like an extended arm of Inner Nørrebro, where everyone is welcome' and 'We must all be here'. But under this surface of the collectively recognised importance of co-existence and fighting spirit were also more pragmatic descriptions, such as when Christian reflected on how 'There is not space for everyone, but everyone has a space'.

Nonetheless, to the regular users and residents of the square, the emphasis was on how the fighting spirit and sense of co-existence was important, particularly in relation to the gangs and police interventions. Our interlocutor, Bent, a man working at the square, told us that he knew that if the gang ever just slightly vandalised the office, he could always go out and talk to them. To him, the young men were potentially dangerous, but on the other hand, co-existence had always proved possible previously.

> I do not like their association [Loyal to Familia]. It has been banned by law and dissolved by law and stuff like that. But on the other hand, we are also human beings and we interact. And that is really important, actually. [. . .] So, that is something we think about. We need to interact. And it may well be that we have a completely different attitude than those out there, and we are against trafficking in hard drugs and crime . . . But it is important to interact anyway.

On one hand, this shows how the square contains all the signs of a place lacking safety, with gangs and police presence, and long-term municipal interventions. On the other, it also shows the square is a place where people live their everyday lives and tolerate each other through notions of co-existence that include the fact that police and gangs are part of the place but are not seen as all-consuming in terms of establishing its atmosphere as inherently 'unsafe'. This taps into criminological studies (Scherg, 2018) in Denmark on safety, *tryghed*, which notes that safety is a mental state of calmness and absence of anxiety. It is not a 'particularly intense experience but more a quiet background emotion . . . as a "state of carefreeness" ' (Scherg, 2018: 3). The gangs and police are there at the margins of attention, like the light that you do not really notice. At least until it is somehow wrong.

Reversing the power of surveillance lighting

The history of protests and, in recent years, intensive gang presence have led the police and city planners to see light as an essential component, although not the sole one, in the fight against crime and feelings of unsafety in the area. Both the police and the municipality would argue, for instance, for installing security lighting aimed to make users feel at ease and create a sense of security (Figures 3 and 6). In practice, this is basically bright, focused light on the wall above two corners of the square where the gang and youths are located at one end and above another entrance to the square by a pub at the other. As the police officers bluntly noted, 'this light would create a feeling of being watched'. We see here an interesting tension between seeing light as an affective force shaping a sense of safety with the users versus being a deterrent and a surveillance instrument for the police and gangs. Yet one of our interviewees, Martin, who had spent most of the last 15 years working at and being on the square, noted this:

> It is a light that draws attention to who is located where, and the ability to see the faces on a camera [. . .] It is meant as a policing tool, as clearly 'safety' can be conceived in different ways. It is not cosy light [*hyggelys*], so to speak.

The lighting, in other words, more than installing a state of carefreeness, may be more about having surveillance and making people and places visible. The reference to cosy light is a cultural preference in Denmark for dimmed lighting, which essentially is about supporting and sensing an intimate, safe and cosy atmosphere of carefreeness (Bille, 2019), detrimental to the bright light that increases security through visibility. 'Safety' and 'security' are, thus, seen more as visibility than as the embodied feeling of a 'positive sense of sheltered-ness, nested-ness, and well-being' (Hutta, 2009: 252; cf. Scherg, 2018).

The instalment of security lights reflects the common perception that *more* light leads to safety (Peña-García et al., 2015). While it may be so in some cases, as Welsh and Farrington (2008) show, the premise is also that quantity of light is the means (and goal), rather than the culturally valued experience of the quality of light. Furthermore, *more* light is not simply about visibility but is involved in local meaning-making processes. Martin, for instance, related how the local fighting spirit and notion of co-existence also meant that there was scepticism about this kind of external intervention in urban design, which showed a focus solely on policing: 'People were tired of the crime but despite this went to the barricades and said "no" to change and the security lighting'. Security lighting – in Danish *tryghedslys*, which has closer reference to the sense of carefreeness – was installed despite such scepticism. It came, however, with subsequent occasional vandalism to the fixtures yet points towards the importance of understanding local perceptions and values in a square, such as a 'fighting spirit'. The residents wanted to decide what to do with 'their' square and came together across political viewpoints and social classes to fight what some saw as a common enemy: the Municipality – at least as long as it did not more wholeheartedly engage the residents, users and organisations. This demonstrates the importance of understanding particular social histories in terms of potential animosity against municipal top-down planning of urban lighting.

Yet the added security lighting did not deter young men from congregating. Rather ironically, the young men have adopted the light. They have not moved elsewhere, nor do they stand in the dark patches or corners of the square but rather straight under the light fixtures located above the pizzeria. That is, rather than preventing crime or congregation through surveillance, the security light attracts the young men to congregate essentially marking their presence, even if such an image of 'controlling' the square may not be their intention. But then again, there is little crime here. As Martin noted, the fact that

Figure 7.6
Security light source
on the wall above the
street.
Photo: Mikkel Bille.

there is so little crime in the square itself 'can be frightening, thinking about what they do in other places in Nørrebro, but they do not shit where they eat'. The surveillance light, intended to create a sense of safety, had in essence made the gangs and youths very much a presence.

The implication of gangs congregating on the corner is exemplified by a resident, Anne. She owns one of the refurbished apartments close to where the gangs congregate. She enjoys the area and goes to and from work, commonly using the square and shops, and has created what she feels is a cosy home. When we brought up the topic of light, the atmosphere the presence of gangs evokes quickly came up. To Anne, her time as a resident at the square has taught her to read the square for small signs and markers to be able to assess if there is trouble in the air: 'There are kind of markers when the atmosphere is really bad [. . .] I knew nothing about it when we moved in, it's something you learn'. These atmospheric markers are closely dependent on ability to 'read' the light. To her, the actual functioning of the public electric lighting shows her when to take her kids straight to the apartment and when they can play in the playground. 'If the light is off, you know that the young men have been on the move and they are committing illegal acts', referring to how vandalism of the lighting fixtures outside her window is connected to recent criminal activity in the area. In this way, a malfunctioning light not only hinders vision but also points to the immediacy of crime, flagging the atmosphere on the square for one's attention.

Yet it is not only the occasional malfunctioning light that concerns Anne. She lives in a large, bright apartment with several large windows from which one can look out over the entire square. Yet once darkness falls, you do not stand and look out at the square.

> It's just something I learned after we moved down here. When we lived on the 5th floor, we could look out the window and people on the square could not see us. But now that we are down here, we can look straight down at the criminals standing below on the square. And I have just learned that I should never, never, never stand at those windows facing the square in the evening with the lights on in the apartment. Because then you can see that we are standing and looking. And I know from some other residents in the housing association that you risk that they [the young men] think you are watching them, staring at them, taking pictures of them. And that can have consequences. You do not want to be *that* person who took pictures.

Generally happy with living by and using the square, she is also keenly aware of the other side of the coin of living in a popular area. Taking pictures is not the only thing to look out for. Anne only gave us a quick glance out of the windows facing the square before urging us to hurry into the living room again. This way of using the windows at night included her children, who had been taught not to stand long in the windows after dark. As another respondent also phrased it, 'You do not want to be the police spy'. Her time as a resident of the square has taught Anne what residents can and cannot do on and around the square, but it has also taught her how to behave within the space to feel safe and where she feels her children can be safe. What she and other interviewees illustrate is that although the young men may not in reality be a threat, their presence is duly noticed and people engage, not in policing as in 'eyes on the street' but rather in self-policing by *not* looking towards the place that is lit by the security light. This extends to neither looking out of the windows nor taking photos in or of the square in the direction of the young men. In this sense, although Anne could easily relax and enjoy her apartment, the square and her neighbourhood, the young men hanging out would still be present at the margins of attention and potentially unsettle her general sense of carefreeness.

Accordingly, the notion of 'more light means greater safety' based on a surveillance premise needs some adjustment in relation to Welsh and Farrington's insight concerning the role of social life. First, more light may attract the very people who invoke fear to occupy the streets, making them *more* visible. Second, the premise of surveillance and 'eyes on the street' rests on the idea that people are not afraid to appear to be watching or be seen as potentially watching. Anne felt she became the object of surveillance, rather than 'the eyes on the street', because of her location with surveillance potential when lights were on. And simultaneously, a lack of light, particularly when caused by vandalism, may foster feelings of danger as it signals what might be about to happen.

Thus, even without addressing the quality of the security lighting, we need to consider how lighting for safety and 'eyes on the street' may also be reversing who and what is watched. What is essentially at stake here is not whether Anne is watching or being watched. Rather, it is the sense of space that the loitering group of men in the bright light imposes upon her own domestic space.

Although *physically* distanced a few metres above and behind walls and windows, at times Anne does not feel an *atmospheric* distance from the people who make her feel unsafe, even if they do not threaten or harm her directly. Instead, for Anne, her home *is* part of the gang's atmospheric space, particularly after dark. For lighting designers and city decision-makers, Anne's account reverses the conventional approach to the relationship between lighting design and safety. It also shows how an atmospheric framing of lighting can reveal a much more complex understanding than technical measurements and point to how light is only one out of many elements that shapes how a place feels. To Anne, the presence of the gangs shaped the atmosphere more than the actual electric light that was only lit in the dark hours. However, both the presence of security lighting making the gangs visible and vandalism to the lights on the square had the effect of instantly shifting the atmosphere towards lack of safety, rather than cosy domestic or urban relaxation.

This atmospheric porosity of built spaces was revealed by several of the people we talked to. Experiencing urban space is not about physical distance but about atmospheric immediacy. One day in early spring, we spoke to a young woman waiting for her friend in the square. She had never been to the square before and felt great being there. On the bench under the trees where she sat, she liked the atmosphere of people playing table tennis and another group of people celebrating a birthday. 'I like it here, but over there I would never go'. She nodded her head in the direction of the corner where the young men were located and explained how even if only a few metres away, she did not feel like going or even looking in that direction for a prolonged time. She felt as if she did not belong there and perceived it as unsafe, regardless of the reality of such danger.

This young woman's way of being part of the square shows another way of seeing the square than as a tangible delimited space. Instead, the square can be experienced as 'constituted by multiple atmospheres that touch, contact, and rub up against one another, rather than a single, overarching, or dominant one' (Anderson & Ash, 2015: 39). An intimate conversation on a bench may feel cosy as those involved are caught up in the conversation, while not noticing that heated debates or a police presence in other parts of the square may evoke a sensation of unsafe intensity. A user may, thus, feel how the square takes on a distinct affective character depending on position and personal story, even if distinct situations are spatially proximate and without any clear material borders, aside from the atmospheric character. One may simultaneously, as Sumartojo and Pink note (2019: 11), think *about* atmospheres around the square as researchers (or designers), which allows us to explore the multiple ways atmospheres are made apprehensible to the users not only as passive receivers of a design intention but also as active facilitators. This atmospheric approach of thinking in, about and through also points to a shift from a mono-sensuous focus on visibility to multisensuous felt atmospheric space.

Spaces of co-existence

The square has undoubtedly gained a reputation due to the crime, vandalism and political demonstrations that have been shown by the media. Mia, a young woman living in one of the refurbished apartments, loved the area but also told us how the media coverage of the square reinforced the negative reputation not only in the surrounding area but in the rest of Denmark as well. Mia lived alone in a small apartment that her sister had bought 15 years ago. Mia grew up in a nearby neighbourhood of Copenhagen and worked for the municipality. She never really felt unsafe living in the area surrounding the square, and she recounts particularly one episode on the square that emphasises how the reputation portrayed by the media creates a different understanding of the square than the one she had:

> My sister moved in approximately 15 years ago. At that time my grandmother lived in Jutland [western Denmark], where she grew up. She and her husband were much like: 'No, Blågårds Plads? I cannot believe that you want to stay in Inner Nørrebro.' The media portray the place in one particular way as a dangerous place, where she could not believe her grandchildren could live, due to the violence. I then took my grandmother down to the square a couple of years ago – she has since died – and at that time she suffered from heavy dementia and could not remember all of the stories from the media and so on. I pushed her in her wheelchair on a sunny day full of life in the square and the first thing she said was: 'God, what a wonderful place you live in.'

The negative stories, so to speak, had coloured the grandmother's understanding of the square, and when the dementia caused her to forget the negative stories, the square felt nice and safe. The residents and regular users had a kind of compartmentalisation, where spaces became encapsulated into atmospheric spaces rather than geometrical spaces dominated by the gang and police presence. It may be cosy or vibrant in one place, while only a few metres away, it would feel intensely dangerous with gangs and police. There is a difference between *seeing* or hearing about a place and *being* in it, as Ebbensgaard noted (2019). As the narrative about the square was lost to the grandmother, the square may aesthetically feel good with all its vibrancy in good weather.

The presence of the youths, the security lighting strategy and the way these are felt by residents and users led us to see how the atmospheric qualities of spaces are shaped through lighting, among other things, not necessarily because of what one sees or one's proximity to it – the lights may be out, or one may be inside one's apartment – but by marking out the (potential) presence of the youths at the margins of attention. The different users of the square accept each other and live in relatively peaceful co-existence – a space for all, even if one disagrees or finds other people intimidating. At least until they make their presence known more than usually through vandalism, for instance.

Prior to our first visit, the local municipal government had initiated a 'Safety Makeover' in the area around Blågårds Plads. They wanted an area where inhabitants and visitors wished to linger – also after dark – and that meant going beyond the security light paradigm of surveillance lighting. This led to new lighting designs being created and installed. One on the wall of the library, called *Universe* (Figure 7.7), and in two passages leading from the square to the rest of Nørrebro, a work titled *Tracks* was installed. These two works were an addition to the security light, lamp posts around public benches under a range of trees and light projection of snowflakes onto the square. Rather than spreading a lot of light, the artistic additions work as orientating light in a passage and pictures cast on a street surface. The lighting designs *Tracks* and *Universe* were created by a designer known for her co-creation art interventions, Karoline H. Larsen, and the lighting design firm Light Bureau. They focused on expressing the local values of co-existence in the new lighting additions to the square. The artist told us that the colour of the light changed every 20 seconds. She had picked the colours after focus group interviewees with young adults using the square, a choice she made to highlight the different approaches people had to the square. She emphasised that the light symbolised the neighbourhood, which always had attracted people from every corner of the world. She wished not only to create light based on the history of the square with its residents from the exposed public housing areas but also to do art that 'makes people feel safe and happy'. The installations made use of coloured light, in *Tracks* using tubes of light and in *Universe* using a projection of collaged images that aimed to spur the imagination by hinting at themes.

The lighting engineer doing the technical work of the lighting design also focused on including the square's residents in the process of making the design. This included thinking of involving the youths in the process of creating the light and making sure they would like it and feel included, even if their attendance during the actual process would turn out to be somewhat limited. The lighting engineer had previously experienced that if the inhabitants had a feeling of ownership, they would not vandalise it, in line with Welsh and Farrington's argument – although he took care to create a design that was less easy to vandalise. This comment must be seen in relation to the ongoing vandalism of other lamp posts on the square that had caused the municipality to add more rugged fixtures that could not be as easily manipulated. The artistic interventions, aimed at attracting different people and opinions, focused on issues around co-existence and diversity, leaving at least in some sense major narratives of safety, crime and vandalism in the shadow, although present.

Figure 7.7
Universe by Karoline H. Larsen.
Photo: Light Bureau.

For Mia, the resident introduced earlier, the artistic lighting intervention reminded her how design sometimes gets too much attention as a sort of fix, as if it creates a new city:

> I think it is such a classic move. I do not know if it is the 'area renewal' or who has initiated it to create safety through lighting and create some art, and maybe it has been an engaging process where the residents have been a part of it. And I can sometimes just think: It is very much a planning ideal that the process has probably been bloody good and you have created some security. But how much does it really change? Well, yes, it's all well and good that there is some street art, but maybe the problems are actually in some other places, which as a planner can be difficult to change, because people also have to have a social boost financially and such.

In favour of Welsh and Farrington's notion of a social theory, we can follow Mia and say that the artistic lighting may in and of itself not change the city, but it may change how people look at the city. By setting up a co-creative process, social challenges are raised and made explicit yet integrated into a more formalised and powerful Safety Makeover policy, whereby the municipality shows a desire to solve the problem by involving the local stakeholders. Simultaneously, the act of co-creation labels the area as a place where those types of planning measures are needed, to follow Mia's argument.

While there is much life on the square that has nothing to do with the police or gangs, the young men were nonetheless mentioned in every conversation with users, stakeholders and inhabitants, also because the specific lighting – or its absence – actually made their presence felt, if nothing else, by proxy through the particular lighting installations drawing attention to the potential of crime and unsafety. You know the men are or could be there, you want to look but you do not, at least not in a way that can be interpreted as staring. You may walk past them on your way out of the square, with no actual chance of you being disturbed by them, yet still notice their presence. Exactly as you feel an eerie presence that may not have anything to do with you, but you just notice it at the margins of attention or more actively by taking a different route or bodily gesture – self-policing that taints the atmosphere yet not necessarily with fear. The people at the margins of attention – lit by security lighting in the dark hours – contribute to how the square feels in certain places and to how specific social conditions are created. For the lighting design to be effective, the designers need to understand the impact of the presence of youths but also without over-determining them. Dealing with safety is not necessarily by trying to make it uncomfortable for them by making them visible but also by seeing what narratives of an area that artistic lighting helps shape, as well as how collective values may allow for social processes to unfold that includes these segments. Rather than the lighting making a difference to the city *as such*, it makes people view the city differently and, through this, *feel* different.

Conclusion

There is a multiplicity of uses of the square, and one can claim that the security lighting has made the presence of police and gangs more visible, drawing attention to issues around safety. Or one could have looked solely at the artistic intervention and focused on how co-existence and diversity are central characteristics of the neighbourhood. However, the square is not *one* space but a multiplicity of felt spaces that unfold in different ways at different times, from different positions, yet still shape an overall impression of an atmospheric vibrancy. It is a vibrancy that our interviewees enjoyed and was lauded in *Time Out*'s review, *despite* the presence of police and gangs. Some atmospheres may be more 'contaminating' than others, such as children's football matches that tinge the square with noise and movement, or the relaxed atmosphere on Friday evenings in the summer. Yet the security lighting and the gang members lurk in the background, sneaking into attention in different ways and places, which may not be about physical distance but about an atmospheric potency. The security light had created an inverted panopticon: a space you do not look at or photograph but engage with through tactics of self-policing.

In other words, what we have illustrated is a place characterised by a 'fighting spirit' where inhabitants live side by side with the gangs and youths spreading unease mostly insignificantly at the margins of attention yet nonetheless there. And in that sense, the atmospheric square is a fluid entity, with waves of atmospheric sensations 'that touch, contact, and rub up against one another' (Anderson &

Ash, 2015: 39). As one of the interviewees said, 'The square is an extension of the atmosphere in Nørrebro'. With the diversity of lighting, from instrumental surveillance to artistic co-creation, lighting design may indeed both help make a difference to the city *as such*, making people and things visible or invisible, but it may also be fruitful to engage with how lighting *feels*. This includes how it is made to feel through co-creative processes and engagement with local sentiments, making people think and feel differently about the city.

Acknowledgements

We would like to thank David Pinder, Jeremy Payne-Frank, Shanti Sumartojo and Siri Schwabe for constructive comments on earlier drafts. Our thanks, too, to the interviewees for opening up to talk about urban spaces, lighting and atmosphere. Funding was provided by the Velux foundation [#16998].

Note

1. https://www.timeout.com/coolest-neighbourhoods-in-the-world. Accessed 2 August 2021.

References

Allen, J. (2006). Ambient power: Berlin's Potsdamer Platz and the seductive logic of public spaces. *Urban Studies*, *43*(2 SPEC. ISS.), 441–455.

Anderson, B. (2009). Affective atmospheres. *Emotion, Space and Society*, *2*, 77–81.

Anderson, B., & Ash, J. (2015). Atmospheric methods. In P. Vannini (Ed.), *Non-representational methodologies* (pp. 34–51). Routledge: London.

Bille, M. (2019). *Homely atmospheres and lighting technologies in Denmark: Living with light*. Bloomsbury: London.

Bille, M., Bjerregaard, P., & Sørensen, T. F. (2015). Staging atmospheres: Materiality, culture and the texture of the in-between. *Emotion, Space & Society*, *15*, 31–38.

Böhme, G. (1998). Atmosphere as an aesthetic concept. *Daidalos*, *68*, 112–115.

Böhme, G. (2013). The art of the stage set as a paradigm for an aesthetics of atmospheres. *Ambiances*, 2–8. http://ambiances.revues.org/315

Casciani, D. (2020). *The human and social dimension of urban lightscapes*. Springer: Milano.

Cook, I. R., & Whowell, M. (2011). Visibility and the policing of public space. *Geography Compass*, *5*(8), 610–622.

Degen, M., & Lewis, C. (2019). The changing feel of place: The temporal modalities of atmospheres in Smithfield Market, London. *Cultural Geographies*, *2013*.

Ebbensgaard, C. L. (2019). Making sense of diodes and sodium: Vision, visuality and the everyday experience of infrastructural change. *Geoforum*, *103*, 95–104.

Ebbensgaard, C. L. (2020). Standardised difference: Challenging uniform lighting through standards and regulation. *Urban Studies, 57*(9), 1957–1976.

Edensor, T. (2013). The gloomy city: Rethinking the relationship between light and dark. *Urban Studies*, *52*, 422–438.

Groening, S. (2014). *Cinema beyond territory: Inflight entertainment and atmospheres of globalization.* Palgrave Macmillan: London.

Grönlund, B. (2014). *Tryghed i forhold til kriminalitet.* Områdefornyelsen Nørrebro. Københavns Kommune.

Hutta, J. S. (2009). Geographies of Geborgenheit: Beyond feelings of safety and the fear of crime. *Environment and Planning D-Society & Space*, *27*(2), 251–273.

Norberg-Schulz, C. (1980). *Genius loci: Towards a phenomenology of architecture.* Rizzoli: New York.

Otter, C. (2008). *The Victorian eye: The physiology, sociology, and spatiality of vision.* University of Chicago Press: Chicago.

Peña-García, A., Hurtado, A., & Aguilar-Luzón, M. C. (2015). Impact of public lighting on pedestrians' perception of safety and well-being. *Safety Science*, *78*, 142–148.

Scherg, R. H. (2018). *Utryghed som fænomen. Er man tryg, hvis man ikke er utryg?* Det Kriminal Præventive Råd: København.

Schivelbusch, W. (1987). The policing of street lighting. *Yale French Studies*, *72*, 61–74.

Schroer, S. A., & Schmitt, S. B. (Eds.). (2018). *Exploring atmospheres ethnographically.* Routledge: London.

Spadoni, R. (2020). What is film atmosphere? *Quarterly Review of Film and Video*, *37*(1), 48–75.

Stephens, A. C., Hughes, S. M., Schofield, V., & Sumartojo, S. (2017). Atmospheric memories: Affect and minor politics at the ten-year anniversary of the London bombings. *Emotion, Space and Society*, *23*, 44–51.

Sumartojo, S., & Pink, S. (2019). *Atmospheres and the experiential world.* Routledge: London.

van Liempt, I., & van Aalst, I. (2012). Urban surveillance and the struggle between safe and exciting nightlife districts. *Surveillance and Society*, *9*(3), 280–292. https://doi.org/10.24908/ss.v9i3.4270

Welsh, B., & Farrington, D. (2008). Effects of improved street lighting on crime. *Campbell Systematic Reviews*, *13*.

Chapter 8

LIGHTS OUT? LOWERING URBAN LIGHTING LEVELS AND INCREASING ATMOSPHERE AT A DANISH TRAM STATION

Mette Hvass, Karen Waltorp and Ellen Kathrine Hansen

DOI: 10.4324/9781003182610-8

Mette Hvass et al.

Introduction: lights out?

In recent years, researchers have noted that people have forgotten to appreciate the aesthetic, social and sustainable qualities of darkness (Dunn and Edensor, 2020). Darkness can sharpen our senses, affect the atmosphere and enrich spatial and social interactions in public space during dark hours. In this study, we look at the role of darkness and light in the urban lighting context through an experiment at a tram waiting area in Aarhus, Denmark's second largest city. We examine how the dimmed lighting introduced in the experiment is perceived compared to the existing bright lighting setting. Focusing on the everyday activity of waiting for – and riding on – the tram amidst the urban rhythm of people, traffic and light, we explore the contradiction between the aesthetic values and safety functions of darkness and light.

Nick Dunn and Tim Edensor elaborate how they perceive this contradiction in their book *Rethinking Darkness*. They align with 'the contemporary academic, creative, ecologically inspired and aesthetic reappraisal of darkness that are challenging the long-standing negative associations that have prevailed until recently' (Dunn and Edensor, 2020: 1). We here seek to examine the positive and negative associations and connotations that darkness elicits in an architectural experiment (Rasmussen, 1966; Pallasmaa, 2012, 2014; Hansen and Mullins, 2014) using anthropological methods to understand the embodied sensory experience of light and darkness in an everyday activity (Pink and Sumartojo, 2018; Sumartojo and Pink, 2017; Sumartojo et al., 2019; Edensor and Hughes, 2021; Ebbensgaard and Edensor, 2021; Thibaud, 2011).

The experiment was conducted at a tram station with urban commuters as part of this everyday (waiting and) transport situation there, with light and darkness as experimental elements of this (Figure 8.1). Tram stations are currently being built in the three largest cities in Denmark to improve urban public transportation and connections to the suburbs, to upgrade the infrastructure and to reduce car traffic (Jensen, 2017). In December 2017, the first tram line opened in Aarhus. Today, 2 tram lines and 48 stations connect the city with the countryside. The stations, both side and island platforms, are implemented in the existing urban built milieu, with a strong visual connection to the surroundings and with traffic on each side. The stations

Figure 8.1 Photos of Nørreport tram station illustrating the change of experience of the local space and the surrounding context during transition hours.
Photo: Mette Hvass.

represent a new element in the built urban environment and introduce new routes and rhythms in the dynamic urban context. People come here to wait and then to move on to their destinations.

The stations were designed by Holscher Design, in close relation with Aarhus Light Rail, and represent a minimalistic and functional design aiming to adapt to the existing urban architecture and transportation system. In order to provide appropriate platform illumination, fixtures are mounted in railings, and in the sheltered waiting area, lighting fixtures are integrated in the roof structure. Originally, dimmable fixtures were intended here, but to achieve cost savings, controls for the lighting were never installed. The result is a brightly illuminated sheltered area that, in the dark hours, contrasts sharply with the downwards-directed low fixtures in the railings and the less brightly lit surroundings. We were interested in how people experience the rhythm of the local space and the surrounding context at Nørreport when they arrive at the tram station to wait for the tram. What effects does the lighting level have on their experiences?

In the following, we introduce an experiment with dimmed light at a tram station and provide a description of methods used to observe and attune to the place and gather empirical data *with* participants (Pink, 2015, 2021; Waltorp, 2020). We then go on to present the analysis of participants' experiences of dimmed light at Nørreport station and focus on the four themes: 1) atmosphere in the waiting area, 2) connectedness to context, 3) connectedness to people and 4) experience of people's activities in the mobile situation. We argue that lower lighting levels produce a relaxed atmosphere and create a visual connection between the waiting area and the surrounding urban context. We conclude that comfort and feelings of safety can be increased in the dimmed lighting at the station but underscore that the lighting levels must be balanced according to the light levels in the immediate surroundings, people's activity in the area as well as the social interactions taking place to benefit optimally from the sensory and affective feelings to which light can contribute.

Analytical framework: light and darkness

The Nordic architectural tradition is rooted in a phenomenological approach concerned with how people experience space. Rasmussen (1966) argues that architecture should be understood and experienced with all the senses in his book *Experiencing Architecture*, discussing how space, materials and rhythm must be felt, and emphasising that light has an important role in telling the story of the experience of the built environment (Rasmussen, 1966: 33). The architect Pallasmaa (2014: 230) describes this phenomenological understanding of the quality of an architectural reality as 'a complex multi-sensory fusion of countless factors' and highlights the role of the peripheral vision and how peripheral vision integrates us in space and lead us to spatial and bodily experiences (Pallasmaa, 2012: 15, see also Gibson, 2014).

Lighting is a factor that not only co-constitutes the optical qualities of a seeing in a space, but also very much the experience of the surroundings in the dark hours. Therefore, we should be aware of the balance between lighting in any given local space for visibility and lighting in the immediate surroundings to enable and reflect both the central focus and the peripheral vision. When focusing on the aesthetic values of lighting, Pallasmaa clearly demonstrates his opinion on too much diffuse light by stating that 'homogenous bright light paralyses the imagination in the same way that homogenization of space weakens the experience of being, and wipes away the sense of place' (Pallasmaa, 2012: 46). Pallasmaa continues with an elaboration on the qualities of darkness: 'Deep shadows and darkness are essential, because they dim the sharpness of vision, make depth and distance ambiguous, and invite unconscious peripheral vision and tactile fantasy' (Pallasmaa, 2012: 46).

To gain a better understanding of the relationship between light, darkness and atmosphere in the urban context, we also draw on the phenomenological theories of Gernot Böhme, who describes the ability of light to unify how the urban context 'illuminations are perceived as atmospheres' because 'all of what is seen takes on a tint that turns the diversity of what is seen into a unified whole' (Böhme, 2017: 202). Böhme uses the theatre as an example of how to articulate an aesthetics of atmospheres (Böhme, 2013). In theatre, lighting has a powerful role in creating atmospheres to enrich the story being told, including the story's emotions and rhythm. Lighting brings life to the stage, the actor and the drama. In the same manner, lighting can be designed to support the urban space, the pedestrians and their actions (Hvass and Hansen, 2021b). The ways in which light and darkness can influence atmosphere in the nocturnal urban context have gained increasing awareness in recent years, as mentioned earlier (Dunn and Edensor, 2020; Edensor, 2015a; Edensor and Hughes, 2021; Sumartojo and Pink, 2017, Pink and Sumartojo, 2018; Sumartojo et al., 2019; Thibaud, 2011). Both in permanent light installations and at urban lighting festivals featuring temporary light installations in an urban context, the atmospheric qualities of lighting are being unfolded (Edensor, 2012a, 2015b).

Dunn and Edensor explore the multiple meanings and uses of darkness across time and space, specifically how darkness has been laden with negative attributes throughout history and how, on the contrary, the positive aesthetic and sensory experiences of darkness *can* prevail (Dunn and Edensor, 2020). In contrast to this approach and related insights, increased illuminance is often combined with a higher degree of perceived safety and less crime within research in the quantitative engineering disciplines of outdoor lighting. When doing tests with higher lighting levels in neighbourhoods with high crime rates, robberies, physical assaults and so forth are reduced (Boyce, 2019), and traffic safety can also be improved (Gibbons et al., 2014). These are measurable indicators that clearly demonstrate the benefits of an increased lighting level. Occasionally, though, the higher lighting levels have an unintended effect because the human aspect of how the lighting feels is disregarded. The aim of this project is to achieve a more fine-grained understanding of people's experience of a specific urban space when we dim the light. We do this paying attention to whether and how lower lighting levels influence the experience of 1) atmosphere in the waiting area, 2) connectedness to context, 3) connectedness to people and 4) experience of people's activities in the mobile situation.

We frame the complexity of a public transit situation in relation to mobility in terms of the rhythm of everyday human practice in the waiting area. At the station, everyday patterns of movement and activity are 'characterized by immanent and emergent possibilities as well as repetitive rhythms' (Edensor, 2012b: 14). This layer of movement, rhythms and constant changes due to car and tram traffic also has an influence on the lighting in the space and is an important element in understanding the space. Furthermore, the presence and flow of people are important factors that should be considered when aiming to understand how dimming of the lighting level is perceived. Such co-presence can be defined as simultaneous presence of individuals in the same physical location but not necessarily engaged in face-to-face interaction with each other (Goffman, 1967). As part of the public realm, Lyn H. Lofland describes 'urban settlements in which individuals in co-presence tend to be personally unknown or only categorically known to one another' (Lofland, 1998: 9). The dimming of the lighting should also support conscious and unconscious meetings between people – a social layer of light that should be designed to meet their needs (Slater et al., 2015).

Combining architectural and anthropological methods in an experiment

Limited research has been done that bring architectural and anthropological methods into dialogue in experiments with light. Anthropologists strive to understand people's situated experiences to, perspectives on and attitudes to a given phenomenon. Architects focus on describing the physical, spatial context based on own observations, sketches and analysis of the space to suggest design improvements, with limited evidence of how people actually experience the space.

The architectural investigation of scale, spatial hierarchy and atmosphere (Rasmussen, 1966; Pallasmaa, 2012; Zumthor, 2006) is rooted in the architectural design tradition of experimentation in a real context (Schön, 1991), as 'the practice of architecture demands the resolution of a complex web of problems in arriving at contextually determined decisions' and the architectural experiment is a tool to arrive at these design decisions (Hansen and Mullins, 2014: 614). The architectural experiment is based on analysis, informing the definition of the design criteria before the actual experiment is designed and tested (Hansen and Mullins, 2014), in what might be called a pilot study in anthropological terminology. In this project, two such analyses of elements were conducted by the first author: one in the field and one in the lab. The first analysis (Hvass and Hansen, 2020) consisted of so-called 'systematic architectural registrations' of the complex urban context. These were conducted by the first author, who is an architect by training, to analyse the role of lighting at the tram station and the surrounding urban context. Registrations were specifically conducted through a 'serial vision' study (Cullen, 1997), conducting a serial of architectural photo registrations walking towards the station and capturing the experience of spatial relation between the station and the surroundings. Furthermore, registrations of being at the station were conducted through photos, architectural sketches and interviews with travellers waiting for the tram (Spradley, 1979). The second analysis was an examination of people's experience of lighting levels between a localised space and the immediate surroundings, which was conducted in a lab setting (Hvass et al., 2021a), gathering information for the design of the final architectural experiment at Nørreport station.

Concept sketches presented later in this text have been used as an architectural tool, partly during the data collecting period and partly during the analysis of the results, to illustrate spatial hierarchies. Sketches were produced in a reflection-in-action process (Schön, 1991) and in an architectural problem-solving process to obtain 'back-talk of self-generated sketches' as the architect and researcher Gabriella Goldschmidt describes the outcome of sketching (Goldschmidt, 2003). The practice of drawing has been used in the research process to frame themes between the authors, while in the final text, the sketches aim to evoke and illustrate findings on the role of lighting in urban contexts. Likewise, photos and luminance maps aim to visually demonstrate the difference between the existing and the dimmed lighting setting at the station and in relation to the regained contact to the surroundings when the lights are dimmed.[1]

The anthropological approach used in this study combine traditional ethnographic methods (Spradley, 1979) and multimodal visual and sensory ethnographic methods (Dattatreyan and Marrero-Guillamón, 2019; Pink, 2015, 2021; Gaver et al., 1999; Waltorp, 2020, 2021). Go-along interviews were conducted over one week in the existing lighting setting and one week in the dimmed lighting setting,

This aimed at an understanding of how the dimmed light influenced the atmosphere in the space and participants' experience of connectedness to the context of waiting together with other travellers. Methods of visual and sensory ethnography have been well described in recent work on urban light and darkness experienced while walking (Sumartojo and Pink, 2017; Sumartojo et al., 2019), giving a rich insight into the complex urban 'lit world' by gathering participants' experiences of everyday routes and light (Sumartojo and Pink, 2017), automated light (Pink and Sumartojo, 2018), and the authors' own experiences in light walks (Sumartojo et al., 2019). By investigating the mobile activities inextricably connected to the waiting at the tram station through go-along interviews, the change in lighting level in the architectural experiment was connected to the repetitive rhythms in the mobile situation (Edensor, 2012b; Lefebvre, 2013). This fostered an opportunity to gather experiences of urban lighting 'toward understanding lighting as part of a configuration of things and processes that make up a perceptual environment – in this case, a lit one' (Sumartojo and Pink, 2017: 371). Drawing on both the architectural and anthropological approaches in this search for the human sensory experience of a space – 1) atmosphere, 2) connectedness to context, 3) connectedness people and 4) people's activities in a dim lighting in the complex urban context – we see it as an advantage to work *with* people in an explorative vein rather than having a predefined template with more closed question-answer interview or questionnaires.

The architectural experiment: existing light versus dimmed light

The site: Nørreport tram station and surroundings

Nørreport tram station is on the edge of Aarhus city centre and serves trams heading in the direction of the suburbs or towards Aarhus's main train station. It is located next to the intersection of the busy street Nørrebrogade and the less busy Nørregade. A two-lane road passes each side of the station, and it affords an unobstructed view of the surrounding urban space. From the station, a three-story residential complex is visible on one side of the street (Figure 8.2); on the other, a four-story building with a fitness centre and a convenience shop on the corner, a three-story office building, and a three-story residential property are located.

Figure 8.2 Photos of Nørreport station in existing and dimmed lighting settings with a residential complex in the background.
Photo: Mette Hvass.

The station, thus, acts as an island in the middle of the road with a pedestrian crossing connecting from pavements on each side of Nørrebrogade. It consists of an arrival area with a check-in stand for tickets and a sheltered area with benches in the middle of the station; it is in this latter area that the lighting is dimmed in the experiment. On each side of the sheltered area, railings function as standing furniture that people can lean against while waiting. The sheltered area is illuminated from fixtures installed in the ceiling, which were dimmed in the experiment. The rest of the station is lit by fixtures in standing furniture and railings mounted at hip height. Two lit commercial posters are installed under the shelter, located on each side of a glass wall and dividing the platform in two. The roads on each side of the station are illuminated by street lighting on masts. The experience of the urban lighting at this specific location is characterised by a complex mix of the headlights from cars driving past; the rhythm of the traffic lights that glow in alternating red, yellow and green; the white tram that arrives and reflects the light from the shelter; and the tram's lights that illuminate the tracks when it arrives. This mix of traffic light is combined with the light from the buildings around the station, with lights that are turned on and off depending on the rhythm of the occupants, and a fitness centre and a convenience shop that are constantly lit with a high light level. This detailed description of the lit world reveals the complexity of both static and dynamic light in this site and highlights the importance of balancing light between a lit space and the architectural, social and mobile traffic context.

The existing and dimmed lighting setting
The shelter ceiling light in the waiting area was dimmed using a neutral density filter attached on the fixture front. The filter reduced the lighting level without compromising the quality of the light and the

ability of the light to reproduce colours. The lux levels in the surroundings, measured horizontal on the pavement close to the surrounding building facades, varied between 1 and 6 lux, except in front of the convenience shop, where 80 lux was measured on the pavement in front of the shop. At the station, the lux levels under the shelter were approximately 165 lux in the existing lighting setting and, in the dimmed lighting setting, approximately 20 lux. This led to an approximately 80% lowering of the horizontal measured lux level on the ground (Figures 8.3 and 8.4).

Figure 8.3 Photo of the existing lighting setting in shelter ceiling.
Photo: Mette Hvass.

Figure 8.4 Photo of dimmed lighting setting in shelter ceiling.
Photo: Mette Hvass.

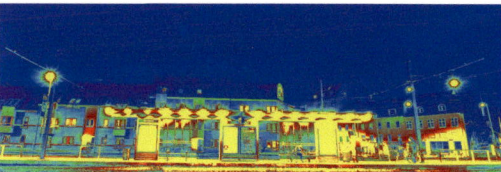

Figure 8.5 Luminance maps, illustrating luminance levels in existing and dimmed lighting settings and how differences in luminance levels impact on the visibility of the surroundings.
Photo: Mette Hvass.

Luminance maps were produced to document the differences in luminance levels on vertical surfaces (Figure 8.5) in the existing and dimmed lighting setting. The luminance maps illustrate how the dimmed lighting setting enables visibility of the surroundings compared to the existing lighting setting. With the dimmed lighting setting, the glass panels under the shelter roof appear transparent and, therefore, the buildings on the other side of the street become visible, as was initially intended in the design of the station. Time-lapses produced during the twilight documented the change in 'the scene' from daylight to electric lighting, capturing how the tram station and the surroundings changed over time – and how the electric lighting 'transforms' the appearance of the space in the dark hours (Figure 8.1).

Research participants
Research participants were recruited via Facebook posts in groups related to Aarhus Tram, through the Consumer Council 'Tænk' and 'Passagerpulsen', and via friends and acquaintances; this yielded ten participants for two interviews each. The group of participants represented different kinds of people, with a mix of commuters and occasional tram users, both women and men, with some in their 20s and others in the mid-50s to late 60s. Prior to the interviews, all participants were informed that they were to experience two different lighting settings, and they were asked to take two to five pictures of a tram station in Aarhus before our first meeting. The recorded interviews were transcribed and coded, sorting the material and manually detecting and recognising themes relating to atmosphere in the space, surroundings, people and their activities.

Mette Hvass et al.

Go-along interviews and participant's photos as probe

The experiment took place in November 2020, a time of year when it is cold and cloudy in Denmark[2]. Participants were interviewed after dark, between 5pm and 8pm. The go-along interview consisted of two meetings: the initial interview was conducted in the existing lighting setting and a subsequent interview in the dimmed lighting setting. The interviewer met each test participant at Nørreport station twice and took the tram to the next stop, Universitetsparken, and then back to Nørreport (Figure 8.6). Walking with other people, with a camera, is a rich sensory ethnographic method, which places the interviewer and the participant on a common ground in the multisensorial activity of moving together and finding a path together (Pink, 2015, 2021; Waltorp, 2020), and this approach draws on both architectural and anthropological methodological traditions. Photos are frequently used in an architectural analysis to get an understanding of space and context (Cullen, 1997). In this case, the architectural interviewer used ethnographic methods to discuss the participants' spatial and social experiences by using their own photos as a tool. The go-along interviews with the tram ride back and forth made it possible to draw nearer to an understanding of the role of light at the station by analysing the series of activities, of which a tram ride consists: arriving at the station, waiting, boarding the tram, taking the actual tram ride, leaving the tram and, finally, leaving the station. Again, with a holistic architectural approach, the interviews enabled examination of the situation from a specific contextual starting point

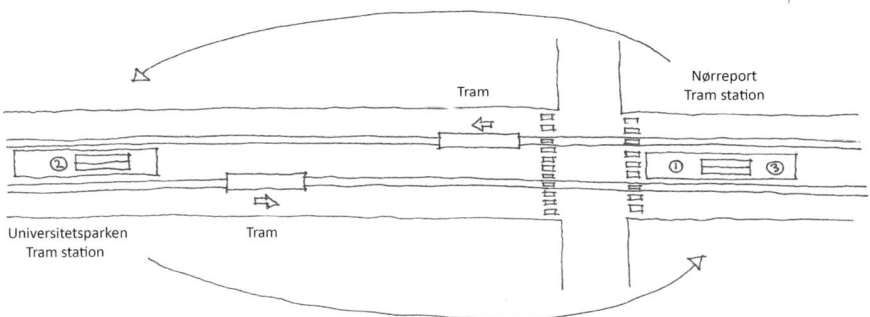

Figure 8.6 Concept sketch illustrating the process of the go-along interview.
Image: Mette Hvass.

(the sheltered waiting area) while moving about with and discussing the participants' sensory experience of the urban context in the two contrasting lighting levels.

Each recorded interview lasted approximately 30 minutes, depending on the arrival and departure time of the tram. The interview was divided in three parts. At arrival, the test participant was interviewed about the experience of the space, people, surroundings and lighting. At the tram station Universitetsparken, we discussed the contrasts to Nørreport, and narratives and memories related to previous transit situations were elicited though people's photographs (Figure 7). When returning to Nørreport, we discussed the 'return' to the station and how this affected the experience of the space. For the second meeting, the three parts of the interview were repeated but with dimmed lighting at Nørreport station. The tram journey was particularly effective, as the differences in the lighting levels between the original lighting setting at the secondary station and the dimmed situation at Nørreport became very clear to the test participants upon returning to Nørreport (Figure 8.7).

Participant-produced images were used throughout the interviews in a number of ways. A first set of photos was produced before the first interview (Pink, 2021: 96). Participants were asked to take photos of a tram station in the dark hours and to capture what their eye caught in the situation (Pink, 2021: 87). A second set of photos was taken during the first interview in the existing lighting setting at Nørreport, at Universitetsparken and again at Nørreport. A third set of photos was taken during the second interview in the dimmed lighting setting at Nørreport, the existing light setting at Universitetsparken and again

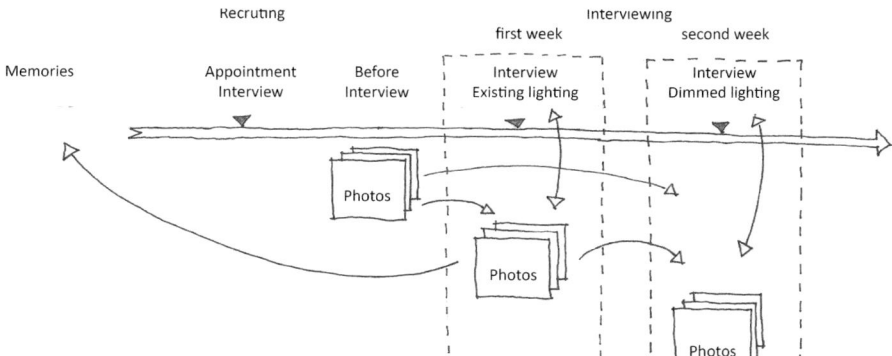

Figure 8.7 Concept sketch illustrating use of photos during interviews.
Image: Mette Hvass.

at Nørreport (Figure 8.6). At each tram stop, a number of questions were posed, and the participant was asked to take photos in relation to the issues discussed. The photos were taken as a collective probe to provoke responses and unexpected ideas (Gaver et al., 1999: 22). The photos led to sudden realisation and triggered memories for the participants (Pink, 2021), opening up to previous experiences with lighting and the sharing of whichever associations this inspired.

Sensing shared space in dimmed light

Light and feelings about light can be difficult to express, indeed because the 'presence of light is often taken for granted in everyday experiences, a vocabulary is lacking' (Sumartojo and Pink, 2017: 1). Participant-produced photos in the process of the go-along interview allowed people to articulate the experienced atmosphere and how they sensed the place and the light. The results from the analysis of the interviews are gathered in four themes, explaining how participants expressed feelings about atmosphere, about connectedness to context and people, and about how the evaluation of the lighting level was linked to people's activities in the mobile situation.

The first theme concerns how the dimmed lighting influenced atmosphere. The majority of the participants expressed a tense atmosphere in the existing lighting setting, where the feeling of being 'exposed' and 'uncomfortable' were dominant, and associations, like 'operating room', 'gas station' and 'prisoner on an island', were used to express how the lighting felt. 'This is calmer and safer, and I feel more protected when I'm not exposed in the light', a participant stated in the dimmed lighting setting. Another participant explained the experience by thoughtfully saying, 'it's nicer now, a more comfortable *room* to be in . . . you're not so illuminated . . . it is something very different . . . it is more intimate'. Some participants linked the dimmed lighting with safety, while others linked the dimmed lighting with a reduced feeling of safety. Two participants in particular felt 'unsafe', 'scared' but also 'fatigued' and 'drowsy' in the dimmed lighting. Others doubted whether their negative feelings about the existing high lighting level were legitimate: 'Even though the lighting level is high and uncomfortable, it's probably the right thing to do to make you feel safe', a participant said, expressing mixed feelings about the uncomfortable high light level and how they supposed that the lighting level was adjusted to meet regulatory requirements for light.

The second theme touched upon how dimming of the light engenders connectedness to the context. The surroundings became visible in the dimmed lighting setting. Some were more positive about this fact than others, but the spatial and social connection to the surrounding space was clear to all. One participant presented new insights after taking photos in the existing lighting setting: 'Well, this wasn't really what I thought I would take a picture of, it is actually quite funny with the contrast. On one side you see the bright tram and on the other side the very dark street. It is quite a big contrast'. This particular person was very focused on safety issues and wished a high lighting level, but while taking a photo, the contrast that the unbalanced lighting levels produced became evident. At the second meeting, in the dimmed lighting, one participant described the visual border that lighting can produce when the contrast between two areas is too high in this way: 'Somehow it seems that the station is better connected to the rest of the city because it is not a light bubble anymore' (Figure 8.8).

Others used words like 'homogeneous' and 'in harmony' to describe the regained contact to the surrounding city: 'Before you stood in quite a lot of light, and then the surroundings became secondary, now it seems more homogeneous'. Some participants did have safety as their highest priority, though, and the feeling of being unsafe in the dimmed lighting was dominant. One participant preferred being exposed in the higher lighting level and felt that the light

Figure 8.8 Concept sketch illustrating the light bubble, with a lack of spatial and social visual interaction due to the high contrast in light level.
Image: Mette Hvass.

generated a safe and enclosed space at the platform. The expressions 'safe enclosed space' or visual disconnecting 'luminous bubble' are in this case contradictory, with a high light level connected to the feeling of safety (Boyce, 2019; Gibbons et al., 2014) on one hand and a high light level connected to the feeling of being exposed and visually disconnected from the context on the other (Pallasmaa, 2012; Hvass et al., 2021a). This example could be regarded as an example of lack of vocabulary (Sumartojo and Pink, 2017: 1) and different understandings of the role of light in everyday experiences. Nevertheless, the regained visual contact with the surrounding context in the dimmed lighting led to increased feelings of belonging and safety for the majority of the interviewed group (Figure 8.9).

The third theme relates to connectedness to people in the waiting area and in the surroundings. One participant contemplated: 'Now you can just see them better [people on the pavement] – see colors, and – just see them. And I feel that I am just as illuminated as they are. . . . it is no longer the case that I am a target, exposed and they can only look at me – now we can just look at each other'. By lowering the light, both spatial and social interactions were improved, both at the station and in relation to the surroundings (Figure 8.9). Participants expressed that they preferred people being present at the platform and

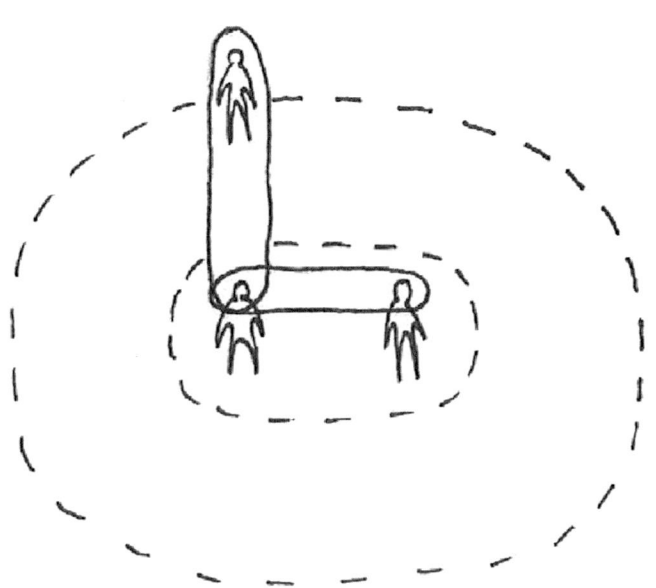

Figure 8.9
Concept sketch illustrating spatial and social interaction in local space and surroundings.
Image: Mette Hvass.

in the surroundings for safety reasons, but they had no wish to interact in the co-presence situation (Lofland, 1998). They came to the platform to wait and wished to isolate themselves, mostly with their mobile telephones. They didn't see it as a criterion that the lighting should support these meetings between strangers. But during the interviews, several participants stated that facial expressions were more relaxed in the dimmed lighting. Commenting on how the lighting effected facial interactions, a participant exemplified this by saying to the interviewer: 'You look more "normal" than last time we met, less scary. Because your skin is a skin color and your glasses look normal, and your jacket is blue and . . . the colors are not blurred'. This indicated that the quality of the lighting has an impact on how we see other people's faces and, thereby, has an influence on the encounters between people in shared public spaces (Goffman, 1967) (Figure 8.10). Ebbensgaard elaborates on visibility in the nocturnal urban space by stating that the role of lighting has to go 'beyond mere provision of visibility and a concern with *what* people see, lighting design is increasingly recognized for its performative effects on *how* people see' (Ebbensgaard, 2020: 1962).

The fourth theme is linked to human activity in the mobile situation. Due to the nature of the go-along interview planned in this field study, the dimmed lighting setting was experienced while waiting, while walking at the station and while departing and arriving with the tram from the nearby station. This led to insights on how the dimmed lighting was experienced as part of a series of consecutive activities. Half of the participants who previously had expressed positive feelings towards the dimmed lighting changed their mind when stepping out of the tram with a high light level into the dimmed sheltered waiting area. Experiencing this transition between the two

Figure 8.10 Concept sketch illustrating light and the public transit situation. *Image: Mette Hvass.*

light zones, a participant exclaimed, 'What a contrast (stepping out of the tram). Crazy. The contrast is greater than when I first came here, where I did not come from a tram. I have to admit, it feels like something's wrong. It feels *very* dark now'. The reaction to the transition in the lighting conditions from the tram to the dimmed shelter exemplifies the fact that the lighting should be designed according to the mobile situation and the repetitive rhythms that the mobile situation consist of (Edensor, 2012b; Lefebvre, 2013) and to the variety of people's behaviour in this specific urban setting (Slater et al., 2015).

Understanding the role of light as part of the contextual whole

This study advances on other explorations of light while moving in the urban context (Pink and Sumartojo, 2018; Sumartojo and Pink, 2017; Sumartojo et al., 2019; Edensor and Hughes, 2021; Ebbensgaard and Edensor, 2021) yet is specific in its design of the architectural experiment. The movement and discussion with research participants take place in a specific context – a station *and* while doing a circular tram ride to the adjacent station and back. The study is an experiment, involving a comparison between how light is perceived before and after dimming in a specific place. By dimming the light, the apparent contradiction between the negative and the positive feeling of darkness can be explored through interviews linked to a specific traffic hub setting where a high light level is normally linked to improved (perceptions of) safety but can also block the view of the surroundings and, therefore, led to diminished feelings of safety.

The experiment is situated both in terms of time, place, surroundings and activity, with waiting for (and riding on) the tram the chosen activities. Our point is that the perception of light and darkness should be evaluated according to a specific activity and with other relevant components taken into consideration rather than 'in the abstract'. By exploring the four themes from the analysis – 1) atmosphere, 2) connectedness to context, 3) connectedness to people and 4) people's activities – we reached a situated understanding of the role of light in the architectural context of Nørreport station in Aarhus. That is, we have shown how a feeling of being a part of the contextual whole can arise when lighting is lowered, allowing less contrast and more balanced lighting levels in comparison to adjacent urban spaces.

The majority of the participants indicated that the atmosphere in the space was more calm, relaxing and intimate in the dimmed light. This is parallel to Pallasmaa's insight (2012: 46) that bright light paralyses the imagination and wipes away the sense of place. Likewise, Brands et al. (2015) explain that bright light reveals potential targets – in this case, the station – and subdues visibility outside the bright area. All participants experienced a regained connection with the surroundings in the dimmed light, including buildings, objects and people passing by. This was perceived as an advantage.

Lighting and security issues in relation to the tram operation and the risk of being hit by a tram were important issues for some, while for others, personal safety and the risk of being assaulted were in focus. The level of the lighting in the experiment was valued to be *too* low – *too* turned down – by the majority of the participants in the go-along interviews. It would have been ideal to have had the opportunity to further test a setting where the light level was less dimmed.

Holscher Design had the intention that their station design should adapt to the existing urban architectural surroundings and transportation system, both during light and dark hours. Due to a reduction in the lighting budget, however, and a lack of knowledge concerning the consequences, the lighting design at the tram station ended up with a continually high lighting level. This design does not adapt to the immediate surroundings in the dark hours as was initially planned and which our experiment shows the importance of. Based on this study, we see potentials for further development of combining architectural and anthropological theory and methods to get closer to an understanding of where and how we can dim lighting around everyday urban activities. Future studies could use the architectural values of light and darkness in the urban spatial hierarchy by programming dynamic lighting settings in relation to the architectural situation and explore how the dimming of light is experienced in relation to other everyday activities and in other contexts.

Conclusion

We know that we need light to see as humans. However, we do not know much about how little light we need to be able to see and at the same time experience the darkness at night, opening up possibilities to see and sense the contexts in which we are located. Our analysis showed that dimmed lighting sharpened the senses and also led to a relaxed atmosphere among the majority of the research participants. Furthermore, dimmed lighting strengthened the experience of connectedness to the surroundings, while the increased spatial and perceptual interaction with the immediate surroundings increased feelings of safety. Some participants, however, felt unsafe in the dimmed light, unsafe in relation to people and unsafe in relation to tram traffic. As the illuminance level on the ground was lowered to approximately 80% of the existing illuminance level, a less drastic dimming could be appropriate in this specific situation.

The results of this research call for attention to the potential in lowering and balancing urban lighting levels between adjacent urban light zones to preserve the atmosphere in an urban space and support improved feelings of safety. Furthermore, by adjusting the lighting to the situation, it is critical to consider the atmosphere, connectedness to context and people, behaviour and the rhythm of people, and traffic in the specific lit space. Participant-produced photos in the collaborative exploration of light paired with interviews conducted while moving were effective in gathering sensory impressions. Such impressions are not as easily elicited in question-answer interviews that are commonly used within engineering and architectural lighting research. This study has drawn on architectural and anthropological approaches by structuring and framing an architectural experiment with two lighting levels (existing and dimmed) and by using ethnographic methods for elicitation of experiences of darkness, light and atmosphere in the tram waiting area and as part of the tram ride. We suggest there is further potential for such a combined approach.

Acknowledgements

We would like to thank our industrial partners Schréder and Holscher Design, AFA JCDecaux for helping to dim the lighting, Aarhus Light Rail for giving us permission to dim the lighting and Keolis for feedback from tram drivers. We would also like to thank the ten engaged test participants who showed up twice for the go-along interviews. And finally, we thank colleagues at the Emerging Technologies Research Lab, Monash University, where Waltorp was a visiting scholar in 2020, and the Lighting Design Research Group at Aalborg University Copenhagen, of which Hansen and Hvass are members.

Notes

1. Concept sketches, photos and luminance maps were produced by the first author.
2. The interviews took place after the first lockdown due to the COVID-19 pandemic in Denmark; the second lockdown was imposed just a few weeks after the experiment was completed. The fact that people had to wear masks and keep their distance had an impact on the results, and participants talked about differences between behaviour at the station before the pandemic and present behaviour, with increased distance-keeping from other travellers.

References

Böhme, G. (2013 October). The Art of the Stage Set as a Paradigm for an Aesthetics of Atmospheres. *Ambiances* 2–8.

Böhme, G. (2017). *The Aesthetics of Atmospheres* (edited by Jean-Paul Thibaud). London: Routledge.

Boyce, P.R. (2019 January). The Benefits of Light at Night. *Building and Environment* 151: 356–367.

Brands, J., Schwanen, T., Van Aalst, I. (2015). Fear of Crime and Affective Ambiguities in the Night-Time Economy. *Urban Studies* 52(3): 439–455.

Cullen, G. (1997). *The Concise Townscape.* London: Architectural Press.

Dattatreyan, E.G., Marrero-Guillamón, I. (2019). Introduction: Multimodal Anthropology and the Politics of Invention. *American Anthropologist* 121(1): 220–228.

Dunn, N., Edensor, T. (2020). *Rethinking Darkness: Cultures, Histories, Practices.* London and New York: Routledge.

Ebbensgaard, C.L. (2020). Standardised Difference: Challenging Uniform Lighting Through Standards and Regulation. *Urban Studies* 57(9): 1957–1976.

Ebbensgaard, C.L., Edensor, T. (2021). Walking with Light and the Discontinuous Experience of Urban Change. *Transactions of the Institute of British Geografers* 46: 378–391.

Edensor, T. (2012a). Illuminated Atmospheres: Anticipating and Reproducing the Flow of Affective Experience in Blackpool. *Environment and Planning D: Society and Space* 30: 1103–1122.

Edensor, T. (2012b). Introduction. In Edensor, T. (Ed.), *Geographies of Rhythm: Nature, Place, Mobilities and Bodies* (1–20). Surrey: Ashgate.

Edensor, T. (2015a). The Gloomy City: Rethinking the Relation between Light and Dark. *Urban Studies* 52(3): 422–438.

Edensor, T. (2015b). Light Design and Atmosphere. *Journal of Visual Communication* 14(3): 331–350.

Edensor, T., Hughes, R. (2021). Moving through a Dappled World: The Aesthetics of Shade and Shadow in Place. *Social & Cultural Geography* 22(9): 1307–1325.

Gaver, B., Dunne, T., Pacenti, E. (1999). Design: Cultural Probes. *Interactions* 6(1): 21–29.

Gibbons, R., Guo, F., Medina, A., Terry, T., Du, J., Lutkevich, P., Corkum, D., Vetere, P. (2014). *Guidelines for the Implementation of Reduced Lighting on Roadways,* Report FHWA HRT 14 050, Federal Highway Administration, Washington, DC.

Gibson, J.J. (2014). *The Ecological Approach to Visual Perception.* London: Routledge.

Goffman, E. (1967). *Interaction Ritual: Essays on Face-to-Face Behavior.* Harmondsworth: Penguin.

Goldschmidt, G. (2003). The Backtalk of Self-Generated Sketches. *Design Issues* 19: 72–88.

Hansen, E.K., Mullins, M. (2014). Lighting Design: Towards a Synthesis of Science, Media Technology and Architecture. *Smart and Responsive Design* (2): 613–620.

Hvass, M., Hansen, E.K. (2020). Architectural and Social Potential of Urban Lighting, a Field Study of How Brightness Can Affect the Experience of Waiting for Public Transportation. *Proceedings PLEA Conference*, September.

Hvass, M., Hansen, E.K. (2021b). Potentials of Light in Urban Spaces Defined through Scenographic Principles. *Proceedings NAF Symposium*, June 2019. In press.

Hvass, M., Wymelenberg, K.V.D., Boring, S., Hansen, E.K. (2021a). Intensity and Ratios of Light Affecting Perception of Space, Co-Presence and Surrounding Context, a Lab Experiment. *Building and Environment* 194: 107680.

Jensen, B. (2017). *Aarhus Letbane, fra vision til virkelighed.* Aarhus: Aarhus Letbane I/S.

Lefebvre, H. (2013). *Rhythm analysis, Space, Time and Everyday Life.* London: Bloomsbury.

Lofland, L.H. (1998). *The Public Realm. Exploring the City's Quintessential Social Territory.* New York: Routledge.

Pallasmaa, J. (2012). *The Eyes of the Skin: Architecture and the Senses.* Chichester, UK: Wiley-Academy.

Pallasmaa, J. (2014). Space, Place and Atmosphere: Emotional and Peripheral Perception in Architectural Experience. *Lebenswelt* 4(1): 230–245.

Pink, S. (2015). *Doing Sensory Ethnography.* London: Sage.

Pink, S. (2021). *Doing Visual Ethnography.* London: Sage.

Pink, S., Sumartojo, S. (2018). The Lit World: Living with Everyday Urban Automation. *Social & Cultural Geography* 19(7): 833–852.

Rasmussen, S.E. (1966). *Om at Opleve Arkitektur.* København: Gads Forlag.

Schön, D. (1991). *The Reflective Practitioner – How Professionals Think in Action.* London and New York: Ashgate Publishing Group.

Slater, D., Sloane, M., Entwistle, J. (2015). *Configuring Light: Staging the Social.* http://www.configuringlight.org/ (Accessed 10 October 2020).

Spradley, J.P. (1979). *The Ethnographic Interview.* New York. Holt: Rinehart and Winston.

Sumartojo, S., Edensor, T., Pink, S. (2019). Atmospheres in Urban Light. *Ambiances* 5(5).

Sumartojo, S., Pink, S. (2017). Moving Through the Lit World: The Emergent Experience of Urban Paths. *Space and Culture* 21(4): 358–374.

Thibaud, J.P. (2011). The Sensory Fabric of Urban Ambiances. *Senses and Society* 6(2): 203–215.

Waltorp, K. (2020). *Why Muslim Women and Smartphones: Mirror Images.* London and New York: Routledge.

Waltorp, K. (2021). Multimodal Sorting: The Flow of Images Across Social Media and Anthropological Analysis. In AB & B. Winthereik (Eds.), *Experimenting with Ethnography: A Companion to Analysis.* Durham: Duke University Press. (Experimental Futures Series).

Zumthor, P. (2006). *Atmospheres. Architectural Environments. Surrounding Objects.* Basel: Birkhäuser.

Chapter 9

TOWERS FOR THE NIGHT

Casper Laing Ebbensgaard

DOI: 10.4324/9781003182610-9

Introduction

While skyscrapers and tall structures historically have been burning their presence onto the night-time sky, crowned with searchlights, advertisement signs and elaborate lighting designs, residential towers remark themselves by their relative architectural absence. They are seldom floodlit, and at night, their structural form tends to recede 'behind' a cacophony of illuminated interiors that give them a granular pattern, which slowly changes with people's movements between rooms, flats and floors. Residential towers project an aesthetic of the vernacular, an expression of the uncoordinated and disorganised life in the vertical night that takes precedence over architectural form and which provides the backdrop for an intimate kind of public to emerge in the everyday spaces of the vertical night (Ebbensgaard, 2020a).

Few have understood the importance of this kind of social patterning at night to the formation of an urban public better than Walter Benjamin. On his late evening and night walks through interwar Naples, he noted how its residential 'buildings rise in tiers . . . as skyscrapers' (Benjamin, 1978: 165), animating the city through their architectural porosity. Pierced by windows, doors, gateways and balconies, and connected through staircases, courtyards and passageways, the Neapolitan skyscrapers 'are used as a popular stage . . . all divided into innumerable, simultaneously animated theatres' (Benjamin, 1978: 167), across which the dramas of everyday life play out for all to see. In recognising the cultural significance of such 'anarchical, embroiled, villagelike' character, Benjamin suggested that the Neapolitan night had more in common with the African *kraal* than it had with Northern European cities not just because of form but because 'the most private affairs' were conducted in communal settings as 'a collective matter' (Benjamin, 1978: 171). At night, intimate life was not restricted to the private sphere but instead constituted a publicly shared sentiment, a shared culture of having the night in common and a shared atmosphere of beholding the vertical night as commons (see also Bille, 2019).

To appreciate the continued importance of the domestic interiors of residential towers in constituting an atmosphere of the night as commons today, one only has to think of the public outcry that results when newly constructed residential towers are left empty and unused. In London, the 50-storey luxury tower St George Wharf became the eye of a public media storm when images of the tower shrouded in darkness started circulating across various social and public media platforms (Booth, 2016). The tower has remained largely unoccupied after completion in 2013, raising concerns about ownership – two-thirds of the flats were bought by foreign investors and a quarter 'held through secretive offshore companies based in tax havens' (Booth & Bengtsson, 2016, para. 3) – and the right of property owners to vacate real estate amidst a national housing crisis. Ownership of property, Ahmed (2019) writes, grants the owner the right not just to use a building but *not* to use it, too; to vacate it; and to squander it, raising questions about the role of private property in the public domain that is considered common. Residential absence, Simone (2014: 57) similarly writes, is a common feature of speculative high-rise development that suggest how these 'projects exist primarily as claims – claims on space that are calculated to posit significant gains only at some future time'. Vacancy of prime real estate brings atmospheric shifts to the vertical city because it no longer teems with the kind of vibrant anarchy that brought Benjamin's Naples to life but instead is defined by the seemingly expanding ghostly residential developments awash with dark, 'dead windows' (Graham, 2016b: 200). The right to determine the conditions under which certain windows are made to bring life or death to the atmosphere of the vertical city – that is, the biopolitics of the vertical night – brings into question the wider politics of orchestrating nocturnal atmospheres in the vertical city.

This chapter is concerned with how residential towers are designed for the night and examines how architects and designers – who adorn the city's skyline with those piercing monuments for elevated inhabitation – consider the nocturnal presence and use of the residential tower as part of shaping their wider urban surrounds. The chapter asks how the design of residential towers for their nightly appreciation and inhabitation offers a platform for considering their potential role in the formation of an urban public night as commons. The chapter addresses the question by briefly examining the 51-storey residential tower Principal Tower, drawing on interview-based materials carried out between 2019 and 2020 with the lead architects, Foster + Partners, the lighting designers Nulty Lighting (interior) and SEAM lighting (exterior), with residents living in the building and with building managers working on them. In the following, I briefly situate the chapter within wider debates on night-time design before discussing the case in three empirical sections that each give attention to different 'surfaces'; the facade design that works as nocturnal camouflage by harnessing and harmonising the building's presence at night; the entrance at ground level where the tower 'meets' the street; and finally, the way the building's lighting system upon initiation started to fade and flicker, in a monumental glitch that reached from the buildings' capillaries into the intimate residential spaces. By attending to the ways light was installed and instilled to hold the building's nocturnal appearance together, the chapter aims at thinking through how tower design might promote more socially and environmentally just designs of the towering night, and how this might contribute to the urban commons.

Dark arts

The recent decades' proliferation of corporately lit towers has made critics argue that the illumination of tall buildings 'produce a rather generic nocturnal text' (Edensor, 2017: 90). With the global advance of easily repeatable formulas and standardised designs, a new species of 'generic skyscrapers' (Easterling, 2014: 12) has normalised the otherwise spectacular form of urban development to the point where iconic high-rise buildings are drowned in vertical forests, reduced to mere 'serial objects of architecture' (Kaika, 2011). As if to resist the pervasive language of dull conformity, elaborate lighting features and integrated designs are being introduced in high-rise buildings in ways that recast skylines as screaming contest where the desperate illumination of 'landmark buildings' seek to justify their presence on the night sky – similarity in diversity.

For residential towers, it is not just their presence on the night sky (the vertical plane) that is marked in light but their insertion into the urban fabric (the horizontal plane) that is cloaked in ever more sophisticated configurations of light and dark, which smooth the otherwise violent insertion of the monstrosity into the urban fabric (Ebbensgaard & Edensor, 2020) – light as lubrication. If we acknowledge the importance of residential towers and the way they appear and inhere at night to the formation of a night as commons, how might we develop an affirmative form of critique (see McCormack, 2012) of the tower block that recognises lighting design's inherent and 'radical potential for challenging and destabilising the appearance of normality'? (Ebbensgaard, 2020b: 1972) If, as Edensor (2017) suggests, light plays a powerful role in redistributing what Jacques Rancière terms the 'field of the sensible', then lighting design holds the potential for destabilising norms of sensation and perception in ways that can challenge the conditions under which certain voices are heard and actions recognised – and we might add, the conditions under which nocturnal atmospheres animate or deaden the urban night.

In tune with this, Nick Dunn suggests that architects and light designers might adopt an ethos of 'Dark Design', in which 'alternative visions for urban places at night' (2020: 24) are developed 'to promote positive, non-consumer-orientated experiences and encounters' (2020: 25). To Dunn, dark design is not simply a question of configuring light and dark elements in a spatial composition – as traditional lighting design might be defined – but a much more radical, adversarial approach to developing alternative experiences, understandings and imaginations of nocturnal environments to mobilise dissent in public discourse, challenging norms, common practice and business as usual. Dark design is as much about visualising spatial 'solutions' as it is about fostering environments in which marginal voices can be heard. To Ebbensgaard and Edensor (2020: 11), such 'alternative experiences' can be drawn from minute instances of sensory disruption in every*night* life as these 'expos[e] the incoherence of the wider urban lightscape' and reveal how 'the city rarely wholly succumbs to . . . homogenising tendencies'. The vibrancy inherent to the urban form shares with the unruly aesthetic of Benjamin's Naples the potential for opening up a space of contestation, where a civic public culture can emerge through continuous contestation over the meanings of urban form, with lighting playing a vital role in '[s]taging social negotiation over meanings of space' (Ebbensgaard, 2015: 120).

To Jane Bennett, this kind of vitality inherent to the material world reveals the powerful agency of matter, suggesting that the 'partition of the sensible' – to stay with Rancière's thought – includes marginalised human *and* non-human 'voices' (2010: vii). With reference to the 2003 blackout on the North American East Coast, she argues that the power outage on the electrical grid was not, as the following public media hunt and federal investigation suggested, due to a single faulty part, personal error or negligence on behalf of an operating body. Rather, the blackout resulted from a cascade of minor, linked events that brought human and non-human agents into a fertile and complex web or interrelations. To Bennett, human or non-human '[e]lements by themselves probably never cause anything' (2010: 33) but always act 'in concert with each other' (2010: 29). In her writings on sensors and their role in making environments programmable, Jennifer Gabrys similarly argues that the capacity for sensation is not restricted to the five senses of 'a founding or original subject' but instead, is a distributed phenomenon 'where everything – even a stone . . . counts as an experiencing subject' (2016: 12–13). The sensors and technologies that are introduced into urban environments under the auspices of making cities 'smarter', by tracking environmental changes and informing human behaviour, predict emergent futures, program urban circulation – and we might add lighting technologies to the mix – themselves become part of the environment they track. Urban technologies are not just artifacts but part of an 'active' world. The point of embracing such a distributed notion of agency is that when considered part of an assemblage constituting, for example, a smart home lighting system or the North American electrical power grid, any one element – human or non-human – is 'incapable of bearing *full* responsibility for their effects' (Bennett, 2010: 37). To Bennett, no single fix can solve a 'problem' like a power outage, and instead, she calls upon 'us' to 'broad[en] the range of places to look for sources' (2010: 37) as any meaningful response to a systematic collapse like the 2003 power outage 'will depend on a host of cooperative efforts' (2010: 30) that reach across material, legislative, social, political and economic domains.

For design critics, or dark designers, this means developing a mode of critique that is less bound to the strictures of surface appearances and open instead to exploring the various relations that emerge through unexpected entanglements. It means that we, as Hayden (1977: 114–115) argues, acknowledge that 'criticizing the design of skyscrapers will not make them disappear', and if we instead explore the relations between 'the perceptions of all skyscraper workers and urban residents, women and men, as well as the specialized insights of architects, artists, and social critics', we might get closer to understanding how they operate – and we might add to the mix the 'voices' of the building, the lighting technologies, the electrons and so on.

Sympathetic towards Bennett's, Gabrys' and Hayden's writings – on lighting, sensors and skyscrapers, respectively – this chapter develops an 'affirmative critique' (McCormack, 2012) of the residential tower as it is designed for nightly appreciation and inhabitation, by recognising the role of lighting design in partitioning and re/distributing the field of the sensible in the vertical night. In the following, I demonstrate how the elaborate lighting scheme for Principal Tower serves as lubrication for vertical development, yet as the elaborate lighting scheme starts to fail, I attend to the systematic glitch not to expose a design flaw but rather to explore the potential for developing a set of propositions for tower block design and development that might promote more socially and environmentally just designs of the towering night.

Facade: *'more* atmospheric'

In the promotional material for Principal Tower, the 51-storey residential tower that was completed in 2018, the chief architects Foster + Partners explain how it is conceived as 'a building of light and shade'. During the day, 'different sections of the building will come alive as sunlight moves around it', reflecting in its shiny, curved glass facade.[1] At night, the flickering rays linger on in its surface, as photovoltaic sun panels placed on the roof power a network of programmable LED strips that have been integrated into every louvre encircling each of its 51 floors, creating an artificial light skin that measures a total of six kilometres. As darkness falls and obscures the tower against the night-time sky, it reemerges as a stack of concentric, glowing circles, all choreographed through the energy flows of recycled light. With diodes placed at 150mm increments, the tower facade has enough 'grain' to create quite spectacular animations but without being an actual media screen.

Initially, Foster + Partners didn't want the light scheme, they 'didn't want it flashy . . . "look at me, look at me" ' and instead proposed – not far removed from Benjamin's vision of a city alive with domestic life – that the building appear only with the following:

> the lighting from the apartments to give it *that* character. Because not all the lighting is going to be on at any one point so we get that quite crumbly aesthetic you get with residential buildings, you know, if it were an office it would be the whole thing lit.

By embracing the pixelated character of an inhabited building, they wanted it to appear 'very understated' and to add 'to that texture' of a city informally lit from the inside and out. The developer was keen to introduce soffit lights on each of the balconies for residents to use throughout the night, but the architect 'feared having one corner with loads of dots – some are on, some are off and we just felt that it wasn't part of the architecture'. Too much texture makes the architecture crumble. They resisted and convinced the developer 'for all these balconies to be lit by the ambient lighting coming from the apartments' in order to preserve 'the informality of people living with different coloured curtains and artwork in the apartments'. Yet the trust that the architect invested in the building's occupants to give the building character throughout the night, but not so much that it would compromise the architecture, did not convince the Canadian developer Concord Pacific that their building could burn itself onto people's retina and cement its place on the London skyline. They were eager to recreate some of the property glitz they have become known for elsewhere (like the ACR Vancouver) and insisted that facade lighting would be 'a good draw for marketing purposes and . . . to sell the remaining apartments'.

Consequently, lighting designers SEAM lighting were brought onto the design team to develop an 'understated' lighting scheme that would be 'respectful of the architecture'. SEAM acknowledged from the start that Principal Tower 'sits amongst . . . all these towers in London . . . that are quite iconic'. Introducing facade lighting would provide an opportunity for the tower to 'demarcate itself' and ensure that while 'it's shouldered up with all these other tall buildings' at night, it would 'be of note'. But rather than altering the building's appearance at day, SEAM proposed to integrate LED strips into the existing fins that run horizontally across each floor so that the lighting would be completely 'seamless with the fin' and, thus, 'could emphasise the curvature of the building'. With the concentric bands in place, SEAM developed a range of animations that could run across the entire facade to manipulate its shape (making it seem slimmer by darkening its corners, making it seem lower by reducing the intensity towards the top) and respond to changes in weather.

Concord Pacific loved these ideas and were eager to know if they could feature 'fireworks' and a big 'countdown for New Year' – much to the dislike of both Foster + Partners and SEAM lighting

Figure 9.1
Principal Tower seen from street level with integrated lighting accentuating the spandrels wrapped around each floor.
Photo: Casper Laing Ebbensgaard.

who instead wanted to create 'a much more sophisticated lighting scheme', 'something simple, something elegant', something like this:

that's *more atmospheric*, so the building could respond to naturally produced phenomena. So, we talked about shimmers and cascades and this kind of moving, inverse floating – like imagine if it snows, you could produce something. Or when it's foggy there could be a

> bit of drifting . . . it could be animated to track wind and gales that come through the City.

Rather than producing a media screen, SEAM's proposed light skin would track changes in the meteorological atmosphere to instigate shifts in the city's affective atmosphere; a computational surface that monitors, measures and computes the environment in order for the building (and its residents) to 'become environmental' (Gabrys, 2016: 16). By claiming to bring its residents and onlooking neighbours into a more intimate relation with 'naturally produced phenomena' – and thus, aligning social and natural worlds through environmental/architectural computation – Principal Tower's light skin is an architectural form of aspirational intimacy with the more-than-human world that its residents come to populate. As SEAM lighting explains, their facade lighting is designed to be as follows:

> very sensitive and respectful of the dark – the dark sky. We have a lot of migrating birds and bats and things like that, so we wanted to keep the light pollution down to a minimum . . . Whilst everything along the street level has got a lot of light – the street lighting, the façade, store front facades – everything that's high up just hasn't. We wanted to blend it into the background so it falls into the night sky.

In this way, the residential high-rise is a mediating technology for nocturnal 'environmental inhabitation' (Gabrys, 2016: 11–12). It promotes ways of inhabiting the night-time sky and dwelling among nocturnal creatures and phenomena that are far removed from the artificial, 24/7 lit city streets below. On the surface, this architectural mediation of social (i.e., artificially lit) and natural worlds (i.e., darkness) sounds like an inversion of what Matthew Gandy has termed 'negative luminescence': the pervasive transformation of the urban ecology so that 'everyday life corresponds increasingly with the abstract calculations of capital rather than the circadian rhythms of life' (Gandy, 2017: 1102). If the tower designed for the night can be more 'sensitive and respectful of the dark' to the extent that it 'falls into the night sky', it is because the light skin is employed as a form of camouflage and, thus, echoes a wider global tendency in which, Stephen Graham argues, towers are 'being powerfully camouflaged' by assertions that their design will contribute to 'increased urban "sustainability"' (Graham, 2016b: 192). Understanding the light skin as a form of nocturnal camouflage is helpful in foregrounding the different ways that it is put to use; on the one hand, it holds the pixelated aesthetic of the building, shaped by incoherent inhabitation together, it is a harness for containing the unruliness of disorganised life within the boundaries of architectural propriety; on the other hand, it softens the towering presence on the night-time sky, making it seem less intrusive, less harmful and more harmonious with its urban and more-than-urban surrounds. The double function of harness and harmony serves as a technology for managing the relations between the private spheres of interior space and the public life of the exterior. But in doing so, the light skin also reveals how the intent of designing the tower for its nocturnal appreciation is at odds with the concern for the lives of the people who will come to inhabit it. As SEAM lighting further explains:

> I always thought that for residential buildings, what's interesting is . . . what's going on in the interior. And so, in a strange way, putting facade lighting on the exterior makes the public see the exterior and pay attention to what's happening on the façade rather than looking into people's houses. It's a weird thing to say, but that way, you can hide something in plain sight, if you just make it a little bit more interesting.

Here, drawing the public's interest to the facade is a way of unseeing what lies beneath it, with surface design acting as a harness that both contains and protects the intimate privacy of high-rise property. This does not, as in the *kraal*, elevate private concerns into a public matter but rather, as with Ahmed's (2019) discussion of use and Simone's (2014) work on real estate speculation, creates the conditions for private ownership and rights to be freed from their obligation towards a wider public good. For all its luminous generosity, spilling into its nocturnal surroundings, Principal Tower is not unduly concerned with the urban atmosphere in which it sits. Instead, it domesticates the human and more-than-human life forms that inhere at night and instrumentalise the atmospheric excess of nocturnal inhabitation as a technology for territorialising the vertical night as property.

Lobby: '*more* bang for your $'

At ground level, where the tower block meets the street, the main entrance and lobby also perform a double function: on the one hand, providing a point of access to the building's core; welcoming residents, visitors and service deliveries; and facilitating a shift in pace, direction and mood when passing across public and private spheres; and on the other hand, regulating *who* and *what* gets to enter, presenting a spatial technology for securitising movement of bodies and keeping them in check under a state of constant, heightened surveillance. It is a rabbit hole, a bottleneck and a portal space that narrows the optics of verticality and imposes metrics that filter movement between horizontal and vertical planes.

Recognising that Principal Tower opens onto part of the city that is busy and renowned for its vibrant nightlife, one of the designers from Foster + Partners explained how they wanted to respect 'that it's someone's home'. While the building has a prominent position within a wider public setting,[2] 'it's not for the general public', and therefore, it is 'not trying to compete' with the surrounding buildings but rather, create a more subdued appearance that shows 'it's for the residents and their visitors'. With the aim of 'making them feel comfortable' and 'welcome', Foster + Partners introduced a flexible lighting system that allowed the building manager to make the lobby 'bright, making it inviting' and ensuring that its 'slightly brighter, maybe warmer' than the surrounding area. This is in order for the lobby to 'feel part of home but separate from your domain, your actual property', and 'that's why we introduced a lot of indirect lighting, to create breaks in the ambience'.

In architectural and urban design discourses, portals are often described as 'gateways' that 'mark the point of transition' (Alexander et al., 1977: 278) between discrete and bounded spaces, and to do so successfully, they must provide an affective shift in the experience of the subject – from one of movement to that of arrival, from that of becoming to that of being (White, 1989: 188). An ambient shift with an existential purpose. In the lobby of Principal Tower, the ambient shift in the passage between the city and the home is instigated through indirectly 'raising the edges, raising the background' of the vertical surfaces in the lobby space. Its vertical illumination also enhances the feel of certain materials, like the veneered timber panels that 'by wall washing them . . . you're getting more bang for your buck because . . . it gives it that additional warmth' and by 'picking up the bumps and swirls and the materials' it contrasts the cold 'and quite hard marble on the floor'. When approaching the lobby, people will see these textured, warm vertical surfaces 'from the out-side in and, once you're inside, you can appreciate that full space . . . it wouldn't feel oppressing'. By wall washing the lobby, the space is opened up, making it 'bright' and 'airy' but without becoming an empty transition space, a void or a portal hole for innocuous passage but rather, a luminous place in its own right, a 'solid thing' (Alexander et al., 1977: 278).

According to the building manager, the lighting, from the external facade to the lobby space, contributes to 'creating [an] atmos-phere and ambience . . . [that] adds something extra', that helps 'make you feel as if you come home to a holiday . . . So you feel relieved and happy to be back, and refreshed when you walk out the door'.

Despite the fully automated controls systems, the manage-ment team need to adjust light levels and temperature according to residents' wishes in order to achieve this experiential surplus: 'the ambient temperature that's comfortable for most residents is about twenty-two Celsius, and it's the same with lights, some people don't like it too bright, some people don't like it too dim'. The constant alteration and adaptation of the ambient infrastructures needs align-ment with changes over the course of the day and with the changing seasons as 'everything is on a sensor and that sensor picks up the amount of sunlight that's outside, and based on that it will either dim or increase the light intensity'. Effectively, it means that the lobby 'is very bright during the day otherwise it just gets lost in the sunlight', but once 'night falls it slowly dims down' for reasons of ambience but also for safety and security reasons 'because the brighter it is inside the lobby the less you can see outside', so they try to make it 'easier for the night staff to see outside and see what happens'.

Managing a luxury building isn't all 'emotional labour'; some-times 'you have to play policeman and just watch things going on', as some residents redecorate their homes, and so the management team make sure they will not 'affect the structure of the building'. Each flat is equipped with a 'home automation system, so you can control everything from your . . . heating, cooling, your lights, under floor heating, towel racks, specific light points in your – in your lounge or bedroom'. And while some residents 'don't like that, they want to go back to basics' the systems are 'connected to some of the [main] facilities and systems in the building . . . [and] nobody apart from the installer can [deinstall] otherwise it might affect the warranty of the

Figure 9.2
Lobby space seen
from the waiting area
for visitors.
*Photo: Casper Laing
Ebbensgaard.*

building'. Yet 'people don't always tell you they're going to do stuff', and the management team have to keep an eye on 'contractors coming into the building and using the lifts with tools and stuff . . . you have to ask the question what are they doing in the apartment'. Even minor decorative work needs to run by the desk:

> A lot of people are installing curtains and, you know, painting walls
> and I have to tell them that they cannot have curtains that don't

> have a cream backing, because then it will affect the external look
> of the building . . . we check every day. I've already told a few
> people off, like, 'don't put blue curtains up, it's in your lease'. . .
> they're not allowed to put up anything that can affect the way the
> building looks . . . like fairy lights on the balcony or things like that.
> They can put coloured lights in the apartments, blue or red or what-
> ever, as long as it's not on the balcony or the external.

The desire for ensuring seamless surfaces that facilitate unhindered circulation of energies, materials and bodies, exposes the work needed to uphold and create what Graham (2016a: 762) terms 'vanity height': a concern not with the quality of living spaces but with their appearance and, thus, creating the high-rise as symbolic value. But of course, everyday life doesn't always slot seamlessly into centralised 'facilities and systems' or play out nicely against a 'cream backing', not even the automated system plays by the rules – as Hayden (1977) reminds us the skyscraper is a wild thing insofar that its material, symbolic and affective qualities elude any attempt to conform – the residential high-rise leaks, seeps and drips with the unruly agency of its human and more-than-human elements. And at night, all these elements seem to erupt through the light in, around and on these buildings.

Golden glitch

Nowhere was the failure to contain the building's buoyancy more evident than when the development in late 2019 neared completion and the first residents started moving in. As they tested the widely anticipated light fantastic, the shimmers and atmospheric shifts quickly started to fade – or flicker, to be precise, as random segments went on and off across different floors creating an uncontrollable splutter of light that refused to conform. A monumental glitch in the atmospheric infrastructure, which, rather ironically, recreated that anarchy that the light skin was meant to harness.

While the residents in Principal Tower spent the better part of 2020 locked inside their flats – and the communal areas, which counted a coffee bar, a swimming pool, a gym and a cinema – engineers and electricians were gliding up and down the facade in a cradle suspended from the roof, 'testing the tubes, individually, to see which ones flicker and which ones work'. By accessing the cavities in between each floor where all the cabling for the main system runs, they 'realised that the cables are too long, and they need to shorten them or replace them with another type of cable'. Like lawless, 'disruptive agents' (Gabrys, 2016: 209) refusing to comply – to perform as prescribed, to act as intended – the tubes exposed the fragility of the modern fantasy of seamless operation and smooth circulation. If we return to Bennett's point about acknowledging the entanglement of human and non-human elements, the glitch cannot be attributed to the failing light rods alone, though, as the effect of such an event always will be 'distinct from the sum of the vital force of each materiality considered alone' (2010: 24). In other words, the glitch is not reducible to an elemental failure and does not expose an inherent design flaw in the singular. The light skin didn't *fail* to instigate the atmospheric shift that was intended to bring its residents into a more intimate relation with the more-than-human elements that inhere in the urban night. Instead, it produced a wholly unexpected atmospheric shift, or shudder, that might reveal how the sensing technology tracked something completely other than what was intended and expected – not the smooth transitions of changing weather but the disruptive and disorienting a/effects of vertical development. To the dark designer, the oddity of seeing a building splutter through the night is, in a way, quite promising because it reminds us

Figure 9.3
Principal Tower light testing, showing several sections faulting or flickering.
Photo: Casper Laing Ebbensgaard.

that these 'seemingly illegitimate contributions . . . challenge us to consider how cities hold together and unfold as sites of political engagement' (Gabrys, 2016: 210). Just like Benjamin's Naples teemed with the unruly vitality of uncoordinated life that escaped the boundaries of architectural propriety – at least according to Benjamin's Northern European customs – this part of Central London is enriched with a wholly unexpected vitalism of electrons running amok in the building's capillaries, dancing to a different beat than the punctual rhythm of capital.

To the residents of Principal Tower, this kind of unpredictability is felt on an intimate plane, as the rebellious charges polluting the facade have infected their automated home lighting systems, piercing the most private spaces of their homes. Take Stephen, who, together with his partner, moved into a quite-tight one-bedroom flat on the 34th floor. The light system that is operational in sections allows them to break up the main living space so that they can create a separation between the otherwise overlapping kitchen, living room, dining room and workstation: 'it's lovely because you can really control it, like to a minute detail'. Via a central computer, they have 'quite granular' control over different sections of the lights that can be preset on four different modes they can switch between by using the 'little touchpads' that are placed around the flat in place of conventional switches: one where the lights are at 70% or 80%, when they are cooking or cleaning; one where it's a bit lower, when they are relaxing in the evening; a third for when they are watching a movie where all the lights are brought down low; and finally, one that turns them all off.

Sometimes, Stephen will adjust the light settings over the course of the evening and night in real time:

> I had friends over for cocktails on Friday and . . . the sun was setting [so] I was like dimming down the lights so you could actually appreciate . . . the sun was setting, you could see the night sky.

The light controls allow Stephen, on the one hand, to accommodate several functions in the same living space at the same time and, on the other hand, to align these uses with shifts in the external environment, allowing 'looking out into a dark city' without being blinded by their own reflection in the floor-to-ceiling windows. Controlling the lighting in conjunction with the changing environment is a form of 'environmental inhabitation' (Gabrys, 2016: 11–12) where shifts in natural and social atmospheres are aligned in the domestic setting. But the lighting is not just about curating the view of the surrounding city:

> because it's a one-bedroom apartment, [guests] have to use our [ensuite] shower room [which is accessed through the bedroom], so I always try . . . to have [the bedroom lights] quite low. And then the bathroom, like usually quite low as well . . . if you were going into somebody's bedroom you don't want to feel like you're being like intrusive.

By controlling the lighting, the intimate spaces of the home that otherwise might have remained hidden to guests are made useful. A lack of domestic space is made bearable through lighting. Yet Stephen's attempt to hold as many functions as possible together in the same space means that moving around the flat often requires brushing up against walls and, evidently, against the touchpads. Advanced as the lighting system is, Stephen quickly discovered this:

> there's a sensor on [the touchpad] . . . and you [move your hand] up to turn it on and down to turn it off. And like . . . every time I walked past the switch, the lights would turn on full or it would turn them completely off. And it was so irritating . . . to walk around the

Figure 9.4
View of Stephen's living room and temporary home working station in adjacent bedroom.
Photo: Casper Laing Ebbensgaard.

apartment and the lights keep turning on and off 'cos you walked past the light switch.

The minor disruptions to the design program are no longer caused by lawless electrons but rather, residents who, when participating in the 'smart' lighting of their homes, themselves become the problem. In considering the challenges of citizen participation, Gabrys suggests that citizens are constructed as problems when 'participation does not unfold according to researchers' plans, but instead irrupts through various idiotic registers that transform the agenda and outcomes of participation' (Gabrys, 2016: 209). If we take the home lighting system, the infrastructure is broken into individual, operational units where each resident has control over their separate parts, and once they fail to operate as intended, when they become dysfunctional, it constructs 'participation as a computable problem' (Gabrys, 2016: 213). In other words, it is not the lighting system that is failing but the way it is computed, managed; and so Stephen 'asked the guys to come back and they turned off that feature because it just was – it was useless'. Yet the night light in the toilet still has a 'life' of its own. It is meant to turn on at the lowest level when a sensor registers the bathroom door opening between midnight and 6am, but on several occasions when Stephen has opened the door in the dead of night, the lights have powered on at the highest level. Even the smart home lighting system evades the attempt to integrate it into the fully computational system of the building.

While the automated home lighting system enables its residents to navigate and negotiate their tight living spaces by curating sequences of ambient atmospheres in continuous concert with the changing meteorological atmospheres, the glitch exposes the limitations of controlling the ambient shifts that the lighting design is intended to provide across urban and domestic spheres. Just as an elaborate facade lighting won't hide empty and unoccupied apartments at night, smart home systems won't make them more liveable – living in the vertical night is an exercise in reckoning with the buoyancy of bouncy electrons that hold seemingly little concern for capital and property. While the lighting design of Principal Tower instrumentalises the excessive vitalism of human and more-than-human life as a technology for territorialising the night, the lively reality of the lighting resists its reduction to serve property ownership.

Figure 9.5
Touchpad to control
the home automation
system in Stephen's
flat.
*Photo: Casper Laing
Ebbensgaard.*

Conclusions

If residential towers behold an aesthetic of the vernacular, it is thanks to the uncoordinated intimacies and disorganised intricacies of domestic life that are not restricted to the private sphere but rather, as Benjamin observed in interwar Naples, constitute a publicly shared sentiment of having the urban night in common. This chapter has explored how the aesthetic quality of vertical inhabitation inheres in the design of residential towers by homing in on the case of Principal Tower – a 51-storey residential tower whose nocturnal appearance and inhabitation is enhanced through elaborate lighting design on the facade, in the lobby and in the domestic interior spaces. By interrogating each of these three luminous planes, the chapter demonstrates how lighting is introduced as an ambient infrastructure for configuring the affective and meteorological atmospheres that are generated in the meeting between the building, its users and its surroundings.

By foregrounding the systematic glitch, the chapter appeals to designers and design critics to take inspiration from these events as an unlikely opening for reimaging the urban night. Rather than exposing their failures as design objects, the lighting infrastructure's resistive operation provides openings for considering how light that spills and seeps across domestic and public spaces become a collective matter of public concern. For the dark designer, it is exactly this transaction between the domestic and city that is key in considering the subtended politics of designing the vertical night; for the dark designer, it is the moments of fissure, failure and flicker that tear away those conventional methods for designating light as good or bad, and instead consider their potential for shaping conditions for collective participation in creating the night as commons.

Notes

1. Retrieved from Principal Tower (2020). https://www.facebook.com/principal tower/videos/1971816099808453/.
2. The immediate surroundings are, in fact, private or, what has become known in London of late as, semi-private spaces; spaces that appear transparently open to use by the general public but which are, in fact, privately managed by security guards and surveillance, who have the right to police and deny any person access without having to give any reason therefore.

References

Ahmed, S. (2019). *Whats the use? On the uses of use*. Durham: Duke University Press.

Alexander, C., Ishikawa, S., Silverstein, M., Jacobson, M., Fiksdahl-King, I., & Angel, S. (1977). *A pattern language: Towns, buildings, construction*. New York: Oxford University Press.

Benjamin, W. (1978). Naples. In P. Demetz (Ed.), *Reflections: Essays, aphorisms, autobiographical writings* (pp. 163–173). New York: Schoken Books.

Bennett, J. (2010). *Vibrant matter: A political ecology of things*. Durham: Duke University Press.

Bille, M. (2019). *Homely atmospheres and lighting technologies in Denmark: Living with light*. London: Bloomsbury.

Booth, R. (2016, May). "Tower for the toffs": UK's tallest skyscraper and playground of the rich. *The Guardian*. https://www.theguardian.com/society/2016/may/24/st-george-wharf-tower-london-luxury-pads-emblematic-housing-crisis

Booth, R., & Bengtsson, H. (2016). The London skyscraper that is a stark symbol of the housing crisis. *The Guardian*. https://www.theguardian.com/society/2016/may/24/revealed-foreign-buyers-own-two-thirds-of-tower-st-george-wharf-london

Dunn, N. (2020). Dark design: A new framework for advocacy and creativity for the nocturnal commons. *International Journal of Design in Society*, *14*(4), 9–23.

Easterling, K. (2014). *Extrastatecraft: The power of infrastructure space*. London and New York: Verso.

Ebbensgaard, C. L. (2015). Illuminights: A Sensory Study of Illuminated Urban Environments in Copenhagen. *Space and Culture*, *18*(2), 112–131. https://doi.org/10.1177/1206331213516910

Ebbensgaard, C. L. (2020a). Light infrastructures and intimate publics in the vertical city. *Urban Geography*, *00*(00), 1–20. https://doi.org/10.1080/02723638.2020.1850001

Ebbensgaard, C. L. (2020b). Standardised difference: Challenging uniform lighting through standards and regulation. *Urban Studies*, *57*(9), 1957–1976. https://doi.org/10.1177/0042098019866568

Ebbensgaard, C. L., & Edensor, T. (2020). Walking with light and the discontinuous experience of urban change. *Transactions of the Institute of British Geographers*. https://doi.org/10.1111/tran.12424

Edensor, T. (2017). *From light to dark: Daylight, illumination and gloom*. Minneapolis: University of Minnesota Press.

Gabrys, J. (2016). *Program earth: Environmental sensing technology and the making of a computational planet*. Minneapolis: University of Minnesota Press.

Gandy, M. (2017). Negative luminescence. *Annals of the American Association of Geographers*, *107*(5), 1090–1107. https://doi.org/10.1080/24694452.2017.1308767

Graham, S. (2016a). Vanity and violence: On the politics of skyscrapers. *City*, *20*(5), 755–771. https://doi.org/10.1080/13604813.2016.1224503

Graham, S. (2016b). *Vertical: The city from satellites to bunkers*. London and New York: Verso.

Hayden, D. (1977). Skyscraper seduction/skyscraper rape. *Heresis*, *2*(1), 108–115.

Kaika, M. (2011). Autistic architecture: The fall of the icon and the rise of the serial object of architecture. *Environment and Planning D: Society and Space*, *29*(6), 968–992. https://doi.org/10.1068/d16110

McCormack, D. (2012). Geography and abstraction: Towards an affirmative critique. *Progress in Human Geography*, *36*(6), 715–734. https://doi.org/10.1177/0309132512437074

Simone, A. (2014). *Jakarta: Drawing the city nearer*. Minneapolis: University of Minnesota Press.

White, E. T. (1989). Path-portal-place. In Matthew Carmona & Steve Tiesdell (Eds.), *Urban design reader* (pp. 185–198). Oxford: Architectural Press.

Chapter 10

DARK DESIGNS

Creating shadow, gloomy spaces and enchanting light

Tim Edensor

DOI: 10.4324/9781003182610-10

Tim Edensor

Introduction

After night falls, while darkness is expected to pervade rural land-scapes, it defamiliarises the daytime city with shifting patterns of illumination and shadow. Light is absorbed, refracted and reflected in very different ways, with impenetrable areas of deep darkness contrasting with bright accumulations of lights, prominent illuminated features and grey washes of shadow. Street lights, empty shops apartment windows and passing cars cast their light along with enticing neon and festive lighting. As Downey (2020: 20) contends, 'centres of power or consumerism grow silent, their doors closed on empty streets' while 'other spaces open, noise and activity move to new zones or overlay diurnal activity centres with alternate uses'. While the nocturnal city is subject to regulatory and highly conventional forms of illumination, it remains a site replete with potential in which the multiple qualities of light design might be more fruitfully deployed. However, in this chapter, I explore both urban and rural contexts to investigate how, rather than focusing on light, we can design with dark.

Until recently, dominant Western conceptions of darkness have been overwhelmingly negative. From medieval times, as Galinier et al. (2010: 820) contend, 'darkness plays an important symbolic role as a metaphor of pagan obscurantism – deviancy, monstrosity, diabolism'. In times of widespread beliefs in the supernatural, the night held multifarious terrors. These superstitious fears were aligned with Christian orthodoxies that underline distinctions between malign darkness and divine light. Most saliently, Genesis 1:2–5 describes how all was subsumed in darkness until God created light. These negative conceptions were further supplemented by Enlightenment ideas that, as Bille and Sørensen (2007: 273) note, signified a process through which scientists might banish ignorance by pursuing 'illumination, objectivity and wisdom'.

As modernity progressed, and under the persistent influence of these ideas, darkness was largely marginalised through what Koslofsky (2011) calls 'nocturnalisation', the expansion of social and economic activity into the night and the subsequent spread of illumination. Most importantly, electrification facilitated the spread of illumination that was conceived as synonymous with the rational circulation of air, light and movement across cities (Schivelbusch,

1988). Subsequently, the expansion of technocratically standardised light design, veering away from earlier urban lighting characterised by 'multiple, overlapping perceptual patterns and practices rather than singular paradigms' (Otter, 2008: 10), has impoverished the aesthetic experience of the city at night and has increasingly extended into non-urban areas.

In the UK, for instance, post-war lighting was deployed at regular intervals, typically sited atop poles to provide a uniform spread of illumination. Shaw (2014) demonstrates how lighting engineers have consistently referred to the British Standards Index BS- 5489:2013 to guide the installation of lighting on roads and in public spaces. Yet these totalising endeavours have rarely resulted in coherent nocturnal urban landscapes, for most urban illumination eventuates from a history of partial infrastructures and succeeding technologies that are supplemented and extended (Dunn, 2019). Yet in drawing on a case study in London, Ebbensgaard (2020: 1959) contends that regulatory constraints can provide frameworks for surprisingly creative, varied light design and preserve darkness; for designers 'translate standards and regulations in ways that challenge the restrictive limitations they pose and resist broader trends to homogenise spatial design', adapting them to environmental, aesthetic and technical conditions and distinct local settings.

Besides the tendency towards illuminated homogeneity, Otter (2008) notes that the unequal distribution of light and darkness have often symbolised urban differentiation. Similarly, Joachim Schlör points out that the brighter the light in the centres, 'the more starkly do the outlines of the darker regions stand out' (1998: 65). Brox (2010: 104) draws attention to how the bright, modern illuminated shop windows, signs, theatre entrances, homes and pubs in city centres have long contrasted with more deprived urban areas, in which 'old light retreated into the far streets and lesser known neighbourhoods, disregarded and disparaged in relation to the new'. Harrison (2015: 14) shows how the highly uneven installation of electric lighting was central to marking racial differences across urban space, with light deemed unnecessary to prevent crime and enhance safety for black residents, thereby naturalising spaces of 'black dispossession and white privilege'. Jaffe et al. (2020) contend that pervasive darkness is also imbricated with sensory perception and, thereby, becomes part of what poverty *feels* like.

These unequal modes of distinction through light design persist (Ebbensgaard and Edensor, 2021). A recent night walk through East London reveals that functional, often glaring lighting in areas of social housing and industry contrast with the modish illumination installed in new upmarket residential developments, which are subtly lit with low-level lighting that, augmented with highlighted trees and architectural features, produces a calm ambience. At the gigantic Canary Wharf financial centre, diverse artistic luminaires play across pavement, steps, lawns and facades, while also glinting in the waters of the repurposed docks. High above, corporate logos blaze from the tops of massive, looming towers. Yet a social lighting design intervention situated in a pedestrian underpass shows how more creative, inclusive urban illumination might re-enchant space.

More seriously, the bright light enshrined in standardisation and resulting from minimal controls imposed on excessive commercial lighting have created vast spaces of over-illumination, with a concomitant excessive expenditure of energy, degraded aesthetics and great harm to human and non-human health. For while darkness is integral to biodiversity, the nocturnal behaviours of breeding, migration, predation and feeding of non-humans have been radically impacted upon by excessive light. Millions of migrating birds are victims of 'fatal light attraction', become disorientated and crash into buildings. The movements of sea turtles, beetles and salamanders on land and at sea are similarly disrupted (Rich and Longcore, 2006), as illumination also extends into a nocturnal marine realm usually conceived as devoid of light. Many non-humans are nocturnal, and artificial lighting exposes them to predators and confuses their biological timing, reducing the time available to find food, shelter and mates. Scientific evidence is also emerging that demonstrates that human circadian rhythms are becoming desynchronised due to the uncoupling of daytime and daylight, night-time and darkness, resulting in disturbed sleep patterns and a rise in obesity, heart disease, hypertension, diabetes, depression and cancer (Rich and Longcore, 2006). These concerns have not been ameliorated by the widespread installation of LED illumination, for the bluer-whiter tonalities of these lights potentially create even greater environmental harm.

In aesthetic terms, over-illuminated designs diminish the human capacity to discern environments by blasting the sensorium with light, reducing the ability to perceive depth and detail, and undermining the gradual visual adaptation to the dark as the eye's rod cells slowly foster visual perception under gloomy conditions. Furthermore, an all-pervasive glare limits the development of the non-visual senses, restricting the sensory and affective experience of nocturnal space (Edensor, 2013). Pallasmaa (2005: 46) explains that encounters with darkness can 'dim the sharpness of vision, make depth and distance ambiguous, and invite unconscious peripheral vision and tactile fantasy', while Serres (2016: 67–68) poetically underlines how darkness solicits non-visual ways of knowing space:

> Night does not anaesthetize the skin but makes it more subtly aware. The body trains itself to seek the road in the middle of darkness, loves small insignificant perceptions: faint calls, imperceptible nuances, rare effluvia, and prefers them to everything loud. Things wandering in the silence and shadow help it to rediscover practices long since lost through forgetfulness and habit.

Over-illumination further limits the experience of the highly diverse stages and forms of gloom. For instance, the three stages of twilight, civil, nautical and astronomical, successively occur during both morning and evening but currently, are rarely discerned. Davidson's (2015) account of the subtle, ever-changing and manifold sensory and metaphorical potency of twilight depicts shifting hues and emerging shadowy forms with thick, poetic descriptions of landscape. Such diversities of darkness also play across urban interior and exterior environments, yet the wash of powerful illumination diminishes our experience of this gloomy diversity.

In addition, over-illumination minimises the potent effects of lighting itself, for light shines most efficaciously when complemented by dark. As Stone (2018) argues, a backdrop of darkness enhances the distinctiveness of the colour, intensity and form of electric lighting, and a more judicious relationship between light and dark promises the development of an environmentally conscious nocturnal sublime. Yet an appreciation of darkness has always co-existed in contestation with the dominant negative assessments identified earlier. For nyctophobia is neither culturally and historically universal, as Nowell and Gonlin (2021), Kumar (2021), Bordin (2021) and Lund (2021) point out in contemporary contexts. Certain religious and spiritual practices have sought the sublimity and mystery of darkness as conducive to meditation (Edensor, 2017), and as William Sharpe notes, a 'second city – with its own geography and its own set of citizens' (2008: 14) emerges when daylight fades. Darkness provides cover for political opposition, illicit romances, subcultural practices and urban exploration, and opportunities for persecuted minorities and marginal groups to escape domineering masters and carve out time and space. At night, urban demi-monde emerges, as witches, prostitutes, bohemians, beatniks, revolutionaries and heretics move through the shadows, and jazz musicians and clubbers come out to play, underpinning darkness's association with libidinal desires and mystery. As I discuss later, darkness is currently undergoing a widespread reevaluation, and design is becoming increasingly enrolled into the production of gloomy and pitch-black spaces.

Designing *with* darkness

In one sense, architects have always worked with darkness, specifically though the manipulation of shadow and shade, ever-present elements of the dark that foreground how light and dark are inextricably relational qualities that pervade the everyday world. Shadows provide texture, depth, perspective and mark boundaries and edges; they banish flat appearances and allow three-dimensional perception. As Kite (2017) discusses, architects from different historical periods have designed buildings with shadow while articulating diverse philosophical and cosmological conceptions about its relationship with light. As Unwin (2020) details, shadow has been manipulated in numerous architectural ways, notably as containers, thresholds, sides and contained shadows, cast shadows, gradients of shade, and views into and out of gloom. The influential 16th-century architect, Andrea Palladio, emphasised perspectival and scenographic effects, for instance, accentuating 'the contrast between the masonry porticos and shadowed recesses' (Hill, 2020: 293). Prominent 19th-century thinker Ruskin (1981: 46) considered that 'I do not believe that ever any building was truly great, unless it had mighty masses, vigorous and deep, of shadow'. And in the 20th century, renowned Norwegian architect Sverre Fehn drew out the nuances of designing with shade, stating that 'each material has its own shadow . . . the shadow of stone is not the same as that of a brittle autumn leaf' (cited in Hill, 2020: 293).

Kite identifies how many architects focused on ecclesiastical interiors, creating dense patches of darkness, supplemented by radiant glow or thin shafts of light, to foster a sense of divine mystery and primeval existence redolent of the cave and the forest. This manipulation of dark and light is evident in the design of crypts, chapels and passages in many medieval gothic churches and cathedrals. Medieval church builders, Tim Ingold (forthcoming) asserts, 'were masters of shadow, of surface convolutions and dark corners, hiding things in alcoves and vaults in such a way that they would appear to emerge only with the shining of the light, through windows or from lamps, only to fade back into the woodwork or masonry once the light had passed'. In this vein, the grand 18th-century baroque spaces devised by Nicholas Hawksmoor and John Vanbrugh and the 19th-century theatrical shadowy spaces designed by John Soane provide great volumes of shadow to the interiors of many of London's iconic buildings. The 19th-century gothic Gorton Monastery in Manchester is similarly designed in accordance with dramatic contrasts between light and shade (see Figure 1). The 18th fantastical etchings of Piranesi's imaginary prison architectures, replete with oppressive areas of spreading gloom in massive interiors, though never realised, have also been influential in undergirding the dramatic potential of darkness in architectural design.

The potency of darkness in architectural design to produce mass, depth, legibility, contrast and mystery has also been eagerly grasped by a range of 20th- and 21st-century architects, chiming with Junichiro Tanizaki's influential celebration of gloomy interiors in his 1933 masterpiece, *In Praise of Shadows* (2001). At a practical level, the production of shadow in sunny climes has proved essential protection for citizens, with the design of verandas, tree-lined streets and covered walkways. In recent years, a further concern with texture and pattern has inspired the design of large facades in which window shades, sills and pilasters are amplified by the sun from shifting angles, reconfiguring surface geometries over the course of the day (Edensor and Hughes, 2021). Most influentially, Kahn (2003) was preoccupied with what he referred to as the 'treasury of shadows', manifest in multiple architectural features, including colonnades, niches, ornamentation and contrasting areas of darkness and light. Scandinavian architects, including Alvar Aalto, Sverre Fehn, Jørn Utzon and Arne Jacobsen, have often been preoccupied with capturing the unique qualities of Nordic light through honouring the play of shadow and light in architectural interiors. Contemporary architects Tado Andao,

Figure 10.1
Manipulation of
dark and shadow
in the gothic
Gorton Monastery,
Manchester.
Photo: Tim Edensor.

Juhani Pallasmaa and Peter Zumthor have sought to infuse the exterior spaces of their buildings with thick rectilinear wedges of shadow. For instance, the deep, moody interior darkness of Zumthor's Bruder Klaus Chapel near Cologne, Germany, is intensified by black charring of the concrete by burning logs.

In addition to ongoing architectural compulsions to produce areas and patterns of shadow, light designers have also increasingly foregrounded darkness as an element integral to design. At a large scale, certain emergent urban policies are devised to manage darkness creatively. For instance, in their Charter on Urban Lighting Further, Lighting Urban Community international (LUCI) assert that cities 'must aim at creating comfortable light environments and protect darker areas' and 'make starlight visible again'. Such perspectives have also been adopted by local authorities; in Eindhoven, a Lighting Master Plan specifies that 'a respect for darkness is a key tenet', and the most immediate effect of this maxim has been a reduction in illuminated advertising. Influential light designer Narboni (2017: 51) has pleaded for a 'nocturnal urbanism' that protects and preserves darkness and supports green spaces and blue areas, such as parks, canals and rivers, by focusing their attention away from illumination. Similarly, lighting design firm Concepto (2012) have developed a master plan for Rennes, France, that introduces *dark zones* into the city core. And as Downey (2021) reports, in Paris, 'public lighting contracts (including street lighting) are set to expire in 2021, entailing a wholesale rethinking of how the city is lit', a process that will seek to establish a different balance between light and dark, where the latter is not conceived as that which must be extinguished. And in the UK, demonstrating how lighting design practice can span both rural and urban settings, Dark Source (2020) have initiated a number of projects to install creative and ecocentric lighting that span dense urban environments, small villages and dark sky parks.

Less mobilised around institutional politics are a group of some 400 light designers who call for the more sustained deployment of what they term the 'Dark Art'. In reclaiming gloom as a positive quality, they declare 'Let there be darkness'. They critique the distorted quantitative measurements that designers and engineers specify as the desirable 'perceived adequacy of Illumination' in particular spaces and, in contrast, promote aesthetic qualities associated with darkness, including mystery, imagination, chiaroscuro, theatricality and atmospheric potency.

Philip Raphael and Chris Lowe belong to this loose affiliation of light designers and outline some of the practical imperatives that designers should undertake in accounting for darkness (2021). In foregrounding the multiplicities that can be produced by exploring the relationship between light and darkness, they emphasise that it is salient to design out light in certain contexts, for rather than simply being additive, they contend, the lighting designer's contribution can be subtractive so as to solicit elements of intimacy, anonymity, mystery and beauty. They discuss how divergent elements of the nocturnal landscape can be highlighted against a dark backcloth to clarify a setting and note that by designing with dimmer levels of light, a greater sensitivity to colour variation, shape and surface is facilitated. Similarly, the deployment of luminaires of varying intensity and tone diversifies how light is refracted, absorbed or reflected by the materialities with which it interacts. They also call for a greater focus on how light and dark can subtly mark the passing of circadian rhythms, changing in response to the darkening and lightening of the day, and show how lighting design can be considered as a sophisticated 'choreographic art', utilised to guide movement through space.

These dark designs were manifest in Lowe and Raphael's experimental event, *Subluminal*, staged in 2014 at Manchester's Victorian, neo-gothic John Rylands Library. In the library's interior, illumination highlighted key architectural features, sculptures, stairs and passages. Entrances and niches shrouded in gloom were complemented by illuminated ornamentation and inscribed surfaces to produce a space of rich contrasts. Here, the reign of light co-existed with dark domains to make the experience of these opposing realms more potent.

New York–based light designer Schwendinger (2021) focuses on how design can be better attuned to the subtle nuances of urban darkness, the diverse 'shades of night' that emerge at different times. Such temporalities accord with the divergent leisure habits and rhythms of work, leisure and consumption, and such need to be acknowledged in the design process, as well as the different affective and sensory experiences that emerge through the night as qualities of light and dark shift. Schwendinger, thus, seeks to incise luminous boundaries into dark volumes, produce double shadows with two separate light sources and create blurred silhouettes on paving and wall surfaces to generate variegated patterns of light and dark. A critical part of her practice is what she terms 'noctambulation', walking through cities at night to become attuned to the diversity of light and dark effects and identify those areas most in need of redesign.

In further promoting the ethos of 'Dark Design', Dunn (2020a: 25) encourages light designers to take on a more adversarial role in local planning, by visualising how cities could be 'designed differently to promote positive, non-consumer-orientated experiences and encounters'. This chimes with Paris-based light designer Matthias Armengaud's contention that buildings have a right to be something other at night. They also, he adds, 'have the right to be nothing at all, to sleep' (cited in Downey, 2020). Dunn also focuses on the distinctive virtues of the urban edgelands where, in contrast to urban centres, there is 'a wider array of dimness and darkness . . . a hinterland between what is known and visible and that which is not' (Dunn, 2020b: 160). In such realms, lighting is less intensively planned, standardised and regulated; often ambient and happenstance; and unconcerned with luring shoppers and leisure seekers. Often characterised by glaring lights and corresponding patches of darkness, edgelands disclose alternative modes of engaging with dark that move away from more homogeneous urban illuminated spaces. This attentiveness to the less traversed nocturnal parts of the city foregrounds how darkness and light are invariably place-specific. In developing their *atNight* project, Barcelona-based architects Martinez-Diez and Santamaria-Varas (2015) have utilised big data to disclose changing patterns of inhabitants' nocturnal use and movement and, thereby, underpin how such activities are highly localised and might inspire a participatory approach to light design.

Alternatively, the preservation of darkness has been sought through the design of responsive and smart lighting that detects light levels or moving bodies and reacts accordingly by being switched on for a limited time. This is exemplified by light designer Sabine de Schutter's lighting scheme for a park in Berlin in which movement patterns and numbers of people are recorded and used to inform an adaptive system of illumination (Edensor, 2017). De Schutter hereby developed the concept of 'Crowd Darkening'; lighting levels rose to enhance feelings of security when few people were in the park, while light levels fell when numbers increased. Besides minimising light pollution, a comfortable public setting in which to socialise was devised.

Crucially, such attempts to design with darkness are not merely deployed by professionals but are utilised in domestic contexts to produce vernacular light designs. Bille (2015) analyses how Danish residents curate the atmospheres and aesthetics of their homes during the hours of darkness to generate *hygge* – a cosy, sociable and affective condition that pervades domestic interior spaces. Candles and low-level luminaires tincture surroundings, blur boundaries between things and produce an enfolding, intimate space amidst the surrounding gloom.

More radical domestic lighting arrangements are practised by Canadian off-gridders who create sustainable living spaces and domestic routines that revolve around a highly responsive engagement with light and darkness. Phillip Vannini and Jonathan Taggart show how these off-gridders negotiate with darkness in a way that those connected to the grid need never consider. On dull days, the batteries they use to collect the power provided by photovoltaic panels will generate only a modicum of power that must be rationed to service the most pressing needs of the household. These homes 'function in greater *synchrony* with the textures of darkness brought on by these natural elements and ephemeral qualities of place', with inhabitants adapting routines and technologies in accordance with the shifting availability of light that depends upon season and weather (2015: 3). Thus, off-gridders must plan ahead and adapt, taking 'darkness into account as both a continuous and repetitive event as well as an indefinite and less foreseeable occurrence' (ibid.: 9), synchronising their everyday rhythms with shifting supplies of sunlight and gloom. Yet this wider experience of darkness is enjoyed; it does not reduce 'visual comfort', for the off-gridders typically enjoy their 'capacity to affect one's immediate environment by way of active participation . . . in one's relative independence from others, and in one's capacity to need and want less' (Vannini and Taggart, 2013).

Besides the contemporary increase in strategies to incorporate darkness into design of nocturnal spaces, there has been a parallel growth in the provision of distinctive attractions that lure visitors by offering a more complete darkness. The appeal of darkness has long traded on the thrills of the supernatural and the uncanny, resonances that persist in contemporary imaginations. Barnaby (2021) demonstrates how, at London's 18th-century pleasure grounds, Vauxhall Gardens, a delight in illuminated spectacles, such as transparent paintings, mirrors and fireworks, were supplemented by dark attractions. In the Hermit's Grotto, Submarine Caves and Dark Walk, visitors could experience a tactile encounter with thick darkness that encouraged excitable social interaction as well as impropriety. Similarly, fairgrounds have deployed darkness to accentuate carnival thrills, constructing tunnels of love, river caves and ghost trains. Dark rides (Zika, 2018) have likewise become a staple feature of many contemporary amusement parks. Visitors are plunged into a hectic whirl of rapid movement and sounds that, in the absence of sight, cannot be anticipated in advance. Carnivalesque walk through dark interior spaces have also been updated with narrow corridors, the effects generated by startling visual technologies and skilful actors deploying darkness to generate delightful frights (Edensor, 2018).

Besides these fairground enchantments, restaurants that offer a dining experience that takes place in complete darkness are increasingly popular (Edensor and Falconer, 2015). Moreover, sites are devised in which music is experienced without any visual distractions, drinking occurs in the absence of light, and plays are staged in total blackness (Welton, 2021). Dancing in the dark events are staged to remove self-consciousness and forestall the potential for women to be subject to a male gaze (Morris, 2021), gay darkrooms permit anonymous sexual exchanges between men, and tourist dark sites, such as *Dialogue in the Dark* offer non-visual sensory encounters with simulations of iconic urban sites (Edensor, 2013). Such venues require careful design to enable the safe passage of visitors and the movements of blind employees who typically act as guides at such attractions. Interior spaces must be sealed from all light, and tactilities, smells and sounds are carefully devised to be stimulating without being overpowering. Comfortable and stable seating, effective acoustics, and smooth pathways are also integral to such designs.

In addition, the deployment of a less that total darkness – yet one that seems complete upon an initial encounter – has proved powerful in the design of certain art experiences. Morris (2011) describes the 2005 event, *The Storr: Unfolding Landscape*, created by the Scottish arts group, NVA, which ran across a spectacular mountainside on the Isle of Skye. The work involved attendees moving across hilly terrain after dark with minimal illumination while encountering diverse light projections, mini-dramas, poetry and music. As a richly 'textured realm of sensory perception' (Morris, 2011: 335), darkness acted to reconfigure apprehension of the landscape.

In Tino Seghal's staged event, *This Variation,* staged in a derelict rail storage depot in Manchester in 2013, visitors entered a room, inside which a thick darkness seemed to pervade, soliciting visitors to imagine its dimensions and shape (Edensor, 2017). In the blackness, chirruping noises and strange voices resounded. However, the room was not as dark as it first appeared, and eyes gradually attuned to the gloom. This gradual perception disclosed that shadowy figures were responsible for the songs and spoken words, and as a romantic, slow song was played, they embraced some visitors in a slow dance. If visitors had entered a fully lit, visible space, they may have cautiously left the scene, yet darkness allowed the dismantling of 'the traditionally passive role of the viewer and the static condition of the artwork' (Metzger, 2013: 4) and the space between performer and audience. Similarly, in her discussion of the *Minamidera* installation by James Turrell on the Japanese island of Naoshima, Sumartojo (2021) shows how the initial experience of darkness devolves into a more attuned visual perception in which the space in front of the viewer is gradually revealed. She explains how Turrell demonstrates that darkness has the capacity to shape space, duration and possibility in ways that elicit new understandings about how we apprehend the world.

Desires for an absence of illumination is notably evident in the proliferating number of places assigned dark sky status, with the International Dark Sky Association having identified a host of preserves, reserves, parks and communities since 2010. Gallaway (2015: 280) summarises the diverse values of the dark night sky as a 'source of aesthetic, scientific and spiritual inspiration . . . a natural resource, a scenic asset and part of humanity's cultural heritage'. As an aesthetic and cultural resource, dark sky areas are haunted by astronomers and ecologists, artists and designers. Stringent lighting codes laid down by the association's Fixture Seal of Approval have inspired design innovations for luminaires that do not contribute to sky glow, light trespass and excessive glare. Techniques of shielding, the minimisation of quantities of blue light, lower colour temperatures, and lower intensity and reconfigured spectral composition of lighting are essential to the gaining of approval (Gaston et al., 2012). As Blair (2016) explores, the small island of Sark in the English Channel has acquired status as a Dark Sky Community by restricting cars and street lighting, while Dark Sky Town Moffat, in South Scotland, has adopted street lighting specifically designed to minimise skyglow.

A non-luminous artistic design is described by Morris (2015), Dalziel + Scullion's 2013 hard plastic benches positioned in a clearing in Galloway's Dark Sky Park amidst the coniferous forest and fashioned with a shallow depression at one end in which visitors might rest their head while looking upwards at the stars. The designers describe the benches as an 'invitation and a conduit to the sensorial resources of hillsides, woodlands, loch sides and riverbanks' of the park, and they are wrought to 'recall the mysterious ancient cup and ring marked stones, and recumbent stones which stand in many parts of Scotland' (Dalziel and Scullion, n.d.). Light designs are also temporarily installed as part of the light festivals that some dark sky parks stage. Galloway Forest Park periodically organises *Sanctuary* (n.d.) in which performances, interactive installations and sound works are supplemented by light projections and radio transmissions, proving opportunities for experimental designs and prototypes. At the Wiston Estate in the South Downs Dark Sky Park, Sussex, England, light works are installed across the gardens, buildings and parkland conjoined by a nocturnal promenade (*Light Up Trails*, n.d.). The event has been subject to some criticism for vanquishing the gloom that such designated realms are intended to provide, with certain bright illuminated displays obliterating the darkness, and yet other, subtler displays stand out against the contrasting blackness. For instance, *Harmonic Portal* by Chris Plant, recently installed at Wiston and previously installed at Durham Lumiere in 1017 (Figure 1), was placed on an ivy-clad wall, a stone wall and a thick hedge. The installation used saturated, ever-changing colours inside and outside a circular frame to intensify a sense of the lithic and vegetal textures of these surfaces, vividly disclosing qualities that may not ordinarily be noticed.

The reconceptualisation of the positive benefits of darkness has triggered the design of environments that are conducive to non-human life as part of the development of a nocturnal commons that is shared by humans and non-humans alike. Dill (2021) points to the installation of cheap, low light barriers to minimise disruption to the breeding cycles of turtles. She also highlights the supremacy of LEDs with their harmful white light, being challenged by low-pressure sodium bulbs that, with their warmly toned, monochromatic yellow glow, minimise skyglow, utilise far less energy and have far less impact on human and non-human circadian rhythms. Critically, the deployment of ecocentric forms of light is not merely a rural concern, for cities, too, offer potentially fruitful settings for non-human flourishing. A path through a London nature reserve in the London borough of Dagenham and Barking has been lit with advanced illumination that does not harmfully impact upon resident bat populations and other sensitive wildlife (INDO, 2021), while the Dutch town of Nieuwkoop has installed an LED lighting with a red glow that is largely imperceptible to the bat's visual system (d'Estries, 2019).

Figure 10.2
Harmonic Portal,
Chris Plant, Durham
Lumiere 2017.
Photo: Tim Edensor.

Dill (2021: 30) calls for the wider adoption of biomimetic, luminescent technologies, for instance, through 'the utilization and cultivation of bioluminescent fungi . . . along footpaths in temperate and other wet climates' and the use of bioluminescent bacteria and algae to avert the need for electric lighting. Glowing lines on the Van Gogh-Roosegaarde bicycle path in the Netherlands inspired by Van Gogh's painting 'Starry Night are composed of ecologically harmless luminescent material that re-enchants nocturnal space while allaying fears about safety (also see Cross, 2019, for a similar example in Poland).

Despite the potential of such design innovations, a greater attunement to the environmental impact of glare on dark-loving non-humans is required to overcome residual practices and conventions in light design (Silver and Hickey, 2020). In engaging with this imperative, Griffiths and Dunn (2020: 213) seek to address the question 'how we can feel our way into an ecosystem, a more-than-human world, through our human sensorium?'. One answer, they suggest, lies in implementing innovative technologies and modes of representation to register the presence of non-humans in dark spaces. They utilise sonographs to pick out the low-frequency sounds made by bats, birds and other non-humans that are indiscernible to humans, drawing attention to a 'hidden polyphony', and create composite photographic images that capture the luminosity of the changing dusk in the North of England, registering infrared and ultraviolet light that is inaccessible to human perception. In fostering the means to 'transgress the limits of our (human) situatedness into those of flora, fauna, and inanimate matter' (ibid.: 217), they enrich an understanding of the night that though it may seem static, silent and passive, in reality, it seethes with multiple, ever-changing non-human energies. In so doing, they have revealed one way in which design might be informed by a much greater awareness of nocturnal vitality.

Conclusion

At present, we live in a vastly over-illuminated world dominated by notions of safety, standardisation, unequally distributed lighting of poor quality and light clutter. The consequences of the ever-encroaching march of electric light into ever more spaces has had a malign effect of human biorhythms and non-human flourishing, with a consequent vast expenditure of energy and contribution to harmful climate change. It is only recently that darkness has been reassessed to the point that suddenly people wish to become reacquainted with the gloom that has largely been lost. These shifting understandings about darkness offer a considerable opportunity for designers to shape the nocturnal world anew, readjusting the relationship between a previously imperial light and a marginalised gloom. As Downey (2020: 16) writes, by 'understanding how articulations of architecture – envelopment, permeability, scale, edge, recess – influence nocturnal spatial practice, alternatives in building and lighting can be imagined'. In this chapter, I have suggested some of the numerous ways in which we can design with shades of darkness ranging from everyday diurnal shadow to sites of complete blackout, and such practices and span urban and rural realms, blurring the reified boundaries between the 'natural' and the 'cultural' that have proved so harmful to greener thinking. Designers are increasingly taking new ecological knowledge, technological innovation and novel aesthetic approaches, diversifying human sensory experience of space and accommodating the needs of our non-human co-inhabitants to place darkness back into the centre of everyday experience. In recalibrating the relationship between light and dark, light itself will gain new potency, since, as Alves (2007: 1254) asserts, darkness 'reaffirms the symbolic power underlying the use of light', and accordingly, 'light only puts on a show if it breaks out of darkness'.

Acknowledgements

I thank Nick Dunn and Shanti Sumartojo for their helpful comments on a draft of this chapter.

References

Alves, T. (2007) 'Art, light and landscape: New agendas for urban development', *European Planning Studies*, 15(9): 1247–1260.

Barnaby, A. (2021) 'In the night garden: Vauxhall Pleasure Gardens 1800–1859', in N. Dunn and T. Edensor (eds) *Rethinking Darkness: Cultures, Histories, Practices*, London: Routledge, pp. 50–60.

Bille, M. (2015) 'Lighting up cosy atmospheres in Denmark', *Emotion, Space and Society*, 15: 56–63.

Bille, M. and Sørensen, T. (2007) 'An anthropology of luminosity: The agency of light', *Journal of Material Culture*, 12(3): 263–284.

Blair, A. (2016) *Sark in the Dark: Wellbeing and Community on the Dark Sky Island of Sark*, St Davids: Sophia Centre Press.

Bordin, G. (2021) 'Inuit's perception of darkness: A singular feature', in N. Dunn and T. Edensor (eds) *Rethinking Darkness: Cultures, Histories, Practices*, London: Routledge, pp. 91–103.

Brox, J. (2010) *Brilliant: The Evolution of Artificial Light*, New York: Houghton Mifflin Harcourt.

Concepto. (2012) *Lighting Master Plan of Rennes, France*. http://www.concepto.fr/portfolio_page/lighting-master-plan-rennes-france/

Cross. (2019) https://www.sustainability-times.com/clean-cities/a-sun-powered-bicycle-path-glows-in-the-dark-in-poland/ (accessed 27/6/2021).

Dalziel and Scullion. (n.d.) *Rosnes Bench*. https://dalzielscullion.com/works-entry/rosnes-bench/ (accessed 3/6/2021).

Dark Source. (2020) https://www.dark-source.com (accessed 27/6/2021).

Davidson, P. (2015) *The Last of the Light: About Twilight*, London: Reaktion.

D'Estries, M. (2019) https://www.treehugger.com/worlds-first-bat-friendly-town-turns-night-red-4868381 (accessed 27/6/2021).

Dill, K. (2021) 'In defense of wild night', *Ethics, Policy and Environment*. doi:10.1080/21550085.2021.1904496.

Downey, C. (2020) 'Shape shifting: Architecture in a wakeful city', in M. Garcia-Ruiz and J. Nofre (eds) *ICNS Proceedings*, Lisbon: ISCTE, pp. 16–29.

Downey, C. (2021) 'Outside knowing: Accessing alterity in the nocturnal urban landscape', in H. Dortdivanlioglu and M. Marratt (eds) *Proceedings of the*

ConCave: Divergence in Architectural Research, Atlanta, GA: Georgia Institute of Technology, pp. 149–157.

Dunn, N. (2019) 'Dark futures: The loss of night in the contemporary city?', *Journal of Energy History/Revue d'Histoire de l'Énergie*, 1(2): 1–17.

Dunn, N. (2020a) 'Dark design: A new framework for advocacy and creativity for the nocturnal commons', *International Journal of Design in Society*, 14(4): 9–23.

Dunn, N. (2020b) 'Urban peripheries as alternative futures', in T. Edensor, A. Kalindides and U. Kothari (eds) *The Routledge Handbook of Place*, London: Routledge.

Ebbensgaard, C.L. (2020) 'Standardised difference: Challenging uniform lighting through standards and regulation', *Urban Studies*, 57(9): 1957–1976.

Ebbensgaard, C.L. and Edensor, T. (2021) 'Walking with light and the discontinuous experience of urban change', *Transactions of the Institute of British Geographers*, 46: 378–391.

Edensor, T. (2013) 'Reconnecting with darkness: Experiencing landscapes and sites of gloom', *Social and Cultural Geography*, 14(4): 446–465.

Edensor, T. (2017) *From Light to Dark. Daylight, Illumination, and Gloom*, Minneapolis, MN: University of Minnesota Press.

Edensor, T. (2018) 'The sensory pleasures of the disoriented tourist', in O. Jensen, S. Kesselring and M. Sheller (eds) *Mobilities and Complexities*, London: Routledge, pp. 56–63.

Edensor, T. and Falconer, E. (2015) 'Dans Le Noir? Eating in the dark: Sensation and conviviality in a lightless place', *Cultural Geographies*, 22(4): 601–618.

Edensor, T. and Hughes, R. (2021) 'Moving through a dappled world: The aesthetics of shade and shadow in place', *Social & Cultural Geography*, 22(9): 1307–1325.

Galinier, J., Becquelin, A., Bordin, G., Fontaine, L., Fourmaux, F., Ponce, J., Salzarulo, P., Simonnot, P., Therrien, M. and Zilli, I. (2010) 'Anthropology of the night: Cross-disciplinary investigations', *Current Anthropology*, 51(6): 819–847.

Gallaway, T. (2015) 'The value of the night sky', in J. Meier, U. Hasenöhrl, K. Krause and M. Pottharst (eds) *Urban Lighting, Light Pollution and Society*, London: Routledge, pp. 267–283.

Gaston, K., Davies, T., Bennie, J. and Hopkins, J. (2012) 'Reducing the ecological consequences of night-time light pollution: Options and developments', *Journal of Applied Ecology*, 49(6): 1256–1266.

Griffiths, R. and Dunn, N. (2020) 'More-than-human nights. Intersecting lived experience and diurnal rhythms in the nocturnal city', in M. Garcia-Ruiz and J. Nofre (eds) *ICNS Proceedings*, Lisbon: ISCTE, pp. 203–220.

Harrison, C. (2015) 'Extending the 'white way': Municipal streetlighting and race, 1900–1930', *Social and Cultural Geography*, 16(8): 950–973.

Hill, J. (2020) 'Architecture in the dark', in M. Butcher and M. O'Shea (eds) *Expanding Fields of Architectural Discourse and Practice*, London: UCL Press.

INDO. (2021) *Bat Friendly Lighting in the Public Realm: A Case Study*. https://indolighting.com/bat-friendly-lighting-project/ (accessed 27/6/2021).

Ingold, T. (forthcoming) 'Commentary 1: On light', in C. Papadopoulos and H. Moyes (eds) *The Oxford Handbook of Light in Archaeology*, Oxford: Oxford University Press.

Jaffe, R., Dürr, E., Jones, G., Angelini, A., Osbourne, A. and Vodopivec, B. (2020) 'What does poverty feel like? Urban inequality and the politics of sensation', *Urban Studies*, 57(5): 1015–1031.

Kahn, L. (2003) *Louis Kahn: Essential Texts*, New York: W. W. Norton.

Kite, S. (2017) *Shadow-Makers: A Cultural History of Shadows in Architecture*, London: Bloomsbury.

Koslofsky, C. (2011) *Evening's Empire: A History of Night in Early Modern Europe*, Cambridge: Cambridge University Press.

Kumar, A. (2021) '*Purda*: The curtain of darkness', in N. Dunn and T. Edensor (eds) *Rethinking Darkness: Cultures, Histories, Practices*, London: Routledge, pp. 79–90.

Light Up Trails. (n.d.) https://www.lightuptrails.com/gallery (accessed 3/6/2021).

Lowe, C. and Raphael, P. (2021) 'Designing with light and darkness', in N. Dunn and T. Edensor (eds) *Rethinking Darkness: Cultures, Histories, Practices*, London: Routledge, pp. 216–228.

Lund, K. (2021) 'Creatures of the night: Bodies, rhythms and Aurora Borealis', in N. Dunn and T. Edensor (eds) *Rethinking Darkness: Cultures, Histories, Practices*, London: Routledge, pp. 127–137.

Martinez-Diez, P. and Santamaria-Varas, M. (2015) 'Atnight project, designing the nocturnal landscape collectively', *51st International Society of City and Regional Planners Congress (ISOCARP): Cities Save the World*, Rotterdam, Netherlands: 19–23 October.

Metzger, C. (2013) *Building Something out of Nothing (or vice versa) in Tino Sehgal's This Progress*. http://www.cylemetzger.com/metzger_sehgal_progress.pdf (accessed 12/5/2021).

Morris, N.J. (2011) 'Night walking: Darkness and sensory perception in a night-time landscape installation', *Cultural Geographies*, 18(3): 315–342.

Morris, N.J. (2015) 'Exhibition review: Dalziel and Scullion, Tumadh: Immersion', *The Senses and Society*, 10(2): 261–266.

Morris, N.J. (2021) 'Dancing in the darkness to *The Darkness*', in N. Dunn and T. Edensor (eds) *Rethinking Darkness: Cultures, Histories, Practices*, London: Routledge, pp. 114–124.

Narboni, R. (2017) 'Imagining the future of the city at night', *Architectural Lighting*. https://www.archlighting.com/projects/imagining-the-future-of-the-city-at-night_o

Nowell, A. and Gonlin, N. (2021) 'Affordances of the night: Work after dark in the ancient world', in N. Dunn and T. Edensor (eds) *Rethinking Darkness: Cultures, Histories, Practices*, London: Routledge, pp. 27–37.

Otter, C. (2008) *The Victorian Eye: A Political History of Light and Vision in Britain, 1800–1910*, Chicago: University of Chicago Press.

Pallasmaa, J. (2005) *The Eyes of the Skin: Architecture and the Senses*, London: Wiley.

Rich, C. and Longcore, T. (eds). (2006) *Ecological Consequences of Artificial Night Lighting*, Washington, DC: Island Press.

Ruskin, R. (1981; first published in 1840) *The Seven Lamps of Architecture,* New York: Farrar, Straus and Giroux.

Sanctuary. (n.d.) https://sanctuarylab.org/about/ (accessed 3/6/2021).

Schivelbusch, W. (1988) *Disenchanted Night: The Industrialisation of Light in the Nineteenth Century*, Oxford: Berg.

Schlör, J. (1998) *Nights in the Big City*, London: Reaktion.

Schwendinger, L. (2021) 'Darkness as canvas', in N. Dunn and T. Edensor (eds) *Rethinking Darkness: Cultures, Histories, Practices*, London: Routledge, pp. 202–215.

Serres, M. (2016) *The Five Senses: A Philosophy of Mingled Bodies*, London: Bloomsbury.

Sharpe, W. (2008) *New York Nocturne: The City after Dark in Literature, Painting and Photography, 1850–1950*, Princeton: Princeton University Press.

Shaw, R. (2014) 'Streetlighting in England and Wales: New technologies and uncertainty in the assemblage of streetlighting infrastructure', *Environment and Planning A*, 46: 2228–2242.

Silver, D. and Hickey, G. (2020) 'Managing light pollution through dark sky areas: Learning from the world's first dark sky preserve', *Journal of Environmental Planning and Management*, 63(14): 2627–2645.

Stone, T. (2018) 'Re-envisioning the nocturnal sublime: On the ethics and aesthetics of nighttime lighting', *Topoi*, 1(1). https://doi.org/10.1007/s11245-018-9562-4.

Sumartojo, S. (2021) 'On darkness, duration and possibility', in N. Dunn and T. Edensor (eds) *Rethinking Darkness: Cultures, Histories, Practices*, London: Routledge, pp. 192–201.

Tanizaki, J. (2001) *In Praise of Shadows*, London: Vintage Classics.

Unwin, R. (2020) *Shadow: The Architectural Power of Withholding Light*, London: Routledge.

Vannini, P. and Taggart, J. (2013) 'Domestic lighting and the off-grid quest for visual comfort'. *Environment and Planning D: Society and Space*, 31(6): 1076–1090.

Vannini, P. and Taggart, J. (2015) 'Solar energy, bad weather days, and the temporalities of slower homes'. *Cultural Geographies*, 22(4): 637–657.

Welton, M. (2021) 'Going dark: The theatrical legacy of Battersea Arts Cantre's Playing in the Dark season', in N. Dunn and T. Edensor (eds) *Rethinking Darkness: Cultures, Histories, Practices*, London: Routledge, pp. 179–191.

Zika, J. (2018) Dark rides and the evolution of immersive media. *Journal of Themed Experience and Attractions Studies*, 1(1): 54–60.

Index

Page numbers in *italics* indicate a figure on the corresponding page.

Index